The Chemical Engineer's Handbook: From Principles to Practice

Welcome to "The Chemical Engineer's Handbook: From Principles to Practice." This comprehensive guide is designed to provide aspiring and practicing chemical engineers with a wealth of knowledge, practical insights, and problem-solving techniques to navigate the exciting and dynamic field of chemical engineering. Whether you are a student seeking to understand the fundamentals or an experienced professional looking to enhance your skills, this book aims to be your trusted companion throughout your journey.

Chemical engineering is a discipline that combines principles from chemistry, physics, biology, and mathematics to solve real-world problems and create innovative solutions. It plays a crucial role in various industries, including pharmaceuticals, petrochemicals, materials, energy, and environmental engineering. The breadth and depth of this field make it both challenging and rewarding, as chemical engineers contribute to advancements that shape our society and improve the quality of life.

In this handbook, we will explore the key principles and concepts that underpin chemical engineering. We will delve into core topics such as thermodynamics, fluid mechanics, heat transfer, mass transfer, reaction kinetics, process design, and control systems. We will also discuss important considerations related to safety, sustainability, and ethics, which are integral to responsible engineering practice.

But this book goes beyond theory. It aims to bridge the gap between academic knowledge and real-world application. Through practical examples, case studies, and problem-solving exercises, you will develop a strong foundation in the fundamental principles of chemical engineering and learn how to apply them to solve complex engineering challenges. Furthermore, you will gain insights from experienced professionals who have successfully navigated the field and understand the intricacies of working in diverse industries.

"The Chemical Engineer's Handbook: From Principles to Practice" is designed to be a comprehensive and accessible resource, suitable for both students and professionals. Whether you are just beginning your journey in chemical engineering or looking to expand your expertise, this book will serve as a valuable reference and guide you through the intricacies of the field.

So, join us on this exciting adventure into the world of chemical engineering. Let's unlock the potential of this dynamic discipline, embrace the challenges, and contribute to shaping a better future through the power of chemical engineering.

Welcome to "The Chemical Engineer's Handbook: From Principles to Practice" – where theory meets practice, and innovation knows no bounds.

Introduction

- Importance of chemical engineering in modern society

Chapter 1: Overview of Chemical Engineering
- Definition and history of chemical engineering
- Interdisciplinary nature of the field
- Role of chemical engineers in various industries

Chapter 2: Fundamentals of Chemical Engineering
- Key principles and concepts in chemical engineering
- Thermodynamics
- Fluid mechanics
- Heat transfer
- Mass transfer
- Reaction kinetics

Chapter 3: Process Design and Analysis
- Process flow diagrams
- Equipment sizing and selection
- Material and energy balances
- Process optimization

Chapter 4: Separation Processes
- Distillation
- Extraction
- Absorption
- Membrane separation

Chapter 5: Reaction Engineering
- Types of chemical reactions
- Reaction rate equations
- Catalysis and catalyst design
- Reactor design and operation

Chapter 6: Process Control and Automation
- Control system fundamentals
- Feedback and feedforward control
- PID controllers
- Process optimization and advanced control strategies

Chapter 7: Safety and Environmental Considerations
- Hazard identification and risk assessment
- Process safety management
- Environmental regulations and sustainability

Chapter 8: Case Studies and Practical Applications
- Real-world examples and case studies
- Problem-solving exercises and simulations
- Application of principles to industrial processes

Chapter 9: Professional Development and Ethical Practices
- Continuing education and professional certifications
- Ethical considerations in chemical engineering
- Professional responsibilities and communication skills

Chapter 10: Emerging Trends and Future Directions
- Advances in technology and materials
- Sustainable and green engineering practices
- Digitalization and Industry 4.0 in chemical engineering

Conclusion
- Recap of key concepts and insights
- Encouragement for further exploration and growth in the field

Appendices
- Reference tables and charts
- Glossary of key terms
- Additional resources and recommended readings

Importance of chemical engineering in modern society

Chemical engineering plays a crucial role in modern society, impacting numerous industries and sectors that are vital to our everyday lives. Here are some key reasons why chemical engineering is important in today's world:

1. Development of Essential Products: Chemical engineering is responsible for the design and development of essential products that we rely on daily. From pharmaceuticals and medical devices to consumer goods and cleaning agents, chemical engineers contribute to the creation of safe and effective products that enhance our quality of life.

2. Energy Production and Sustainability: Chemical engineers play a vital role in the production and optimization of energy resources. They work on developing and improving processes for the extraction, refining, and distribution of fossil fuels. Additionally, chemical engineers are at the forefront of researching and developing renewable energy sources such as solar, wind, and biofuels, contributing to a more sustainable future.

3. Environmental Protection: With growing concerns about climate change and pollution, chemical engineers are actively involved in developing technologies and processes to mitigate environmental impact. They work on projects related to waste management, water treatment, air pollution control, and the reduction of

greenhouse gas emissions.

4. Food and Beverage Production: Chemical engineers contribute to the safe and efficient production of food and beverages. They are involved in areas such as food processing, preservation, and packaging, ensuring that food products meet quality standards and are safe for consumption.

5. Biotechnology and Pharmaceuticals: Chemical engineers are at the forefront of the biotechnology and pharmaceutical industries. They work on developing innovative drugs, vaccines, and therapies, as well as optimizing production processes for these life-saving products.

6. Material Science and Nanotechnology: Chemical engineers contribute to the development and optimization of advanced materials with unique properties. They work on projects involving nanotechnology, polymers, composites, and specialty chemicals, driving advancements in industries such as electronics, aerospace, and automotive.

7. Water and Air Quality: Chemical engineers play a critical role in ensuring access to clean water and improving air quality. They work on projects related to water treatment and purification, wastewater management, and air pollution control technologies.

8. Safety and Process Optimization: Chemical engineers prioritize safety in all aspects of their work. They develop and implement safety protocols to minimize the risk of accidents in chemical processes and industrial facilities. They also optimize processes to improve efficiency, reduce costs, and enhance productivity.

Overall, chemical engineering is indispensable in modern society due to its contributions to essential products, energy production, sustainability, environmental protection, healthcare, and technological advancements. The expertise and innovation of chemical engineers continue to shape and improve our world, making it safer, more efficient, and more sustainable for future generations.

Definition and history of chemical engineering

Chemical engineering is a branch of engineering that combines principles of chemistry, physics, mathematics, and biology to design, develop, and optimize processes and systems for the production, transformation, and utilization of chemicals, materials, and energy. Chemical engineers apply their knowledge and skills to various industries, including manufacturing, pharmaceuticals, energy, environmental protection, food and beverage, and biotechnology.

The history of chemical engineering can be traced back to the Industrial Revolution in the 18th century when advancements in chemical processes and industrial production began to emerge. Early pioneers such as George E. Davis, Lewis M. Norton, and William H. Walker laid the foundation for the discipline by applying scientific principles to industrial processes. In the late 19th and early 20th centuries, chemical engineering education programs were established in universities, and the discipline began to gain recognition as a distinct field of study.

The development of chemical engineering accelerated during World War I and World War II when chemical processes played a critical role in supporting the war efforts. During this time, chemical engineers were involved in the production of explosives, fuels, synthetic materials, and pharmaceuticals.

In the post-war era, chemical engineering expanded its scope to address emerging challenges and societal needs. The discipline contributed to advancements in petroleum refining, petrochemicals, plastics, polymers, and fertilizers, fueling the growth of various industries. With the increasing importance of environmental protection and sustainability, chemical engineers have also been instrumental in developing cleaner and more efficient processes, waste management systems, and renewable energy technologies.

Today, chemical engineering continues to evolve and adapt to the changing needs of society. The field incorporates advancements in areas such as nanotechnology, biotechnology, sustainable energy, and process optimization. Chemical engineers are at the forefront of developing innovative solutions to complex problems, striving to improve efficiency, minimize environmental impact, and enhance safety in industrial processes.

In summary, chemical engineering has a rich history rooted in the Industrial Revolution and has evolved into a diverse discipline that encompasses a wide range of industries and technologies. Its contributions to society have been significant, and chemical engineers continue to play a crucial role in shaping the world by advancing scientific knowledge, improving processes, and addressing global challenges.

Interdisciplinary nature of the field

One of the defining characteristics of chemical engineering is its interdisciplinary nature. Chemical engineers draw knowledge and principles from various scientific and engineering disciplines to solve complex problems and design innovative processes. Here are some key disciplines that intersect with chemical engineering:

1. Chemistry: Chemical engineering is deeply rooted in chemistry, as it deals with the transformation of chemicals and the understanding of chemical reactions. Chemical engineers apply principles of organic, inorganic, physical, and analytical chemistry to design and optimize chemical processes.

2. Physics: Understanding the principles of physics is crucial for chemical engineers, as they deal with phenomena such as heat transfer, fluid mechanics, and thermodynamics. This knowledge helps in designing efficient heat exchangers, optimizing reactor performance, and understanding the behavior of materials.

3. Mathematics: Mathematical modeling and analysis play a significant role in chemical engineering. Chemical engineers use mathematical equations to describe and predict the behavior of systems and processes. They apply concepts from calculus, differential equations, and numerical methods to solve complex problems and optimize process design.

4. Biology: With the rise of biotechnology and bioengineering, knowledge of biology has become increasingly important in chemical engineering.

Chemical engineers work with biological systems, such as enzymes and microorganisms, to develop biofuels, pharmaceuticals, and other bioproducts. They apply principles of biochemistry and molecular biology to optimize processes involving biological materials.

5. Environmental Science: Chemical engineers are concerned with the environmental impact of industrial processes and work to develop sustainable solutions. They collaborate with environmental scientists to ensure compliance with regulations, develop pollution prevention strategies, and design waste treatment and disposal systems.

6. Materials Science and Engineering: Understanding the properties and behavior of materials is essential for chemical engineers, especially in areas such as polymers, composites, and nanomaterials. They study the structure, composition, and properties of materials to design and develop new materials with specific properties.

7. Process Engineering: Chemical engineers apply principles of engineering, such as process design, optimization, and control, to develop efficient and cost-effective manufacturing processes. They consider factors such as plant layout, equipment selection, safety, and economics to design and operate chemical processes.

The interdisciplinary nature of chemical engineering allows for a holistic approach to problem-solving and innovation. Chemical engineers integrate knowledge from these various disciplines to tackle complex challenges in industries such as energy, pharmaceuticals, food production, environmental protection, and more. Their ability to combine expertise from multiple fields makes chemical engineering a versatile and dynamic discipline that plays a vital role in improving processes, developing new technologies, and addressing global challenges.

Role of chemical engineers
in various industries

Chemical engineers play a vital role in various industries, using their expertise to design, develop, and optimize processes and systems that involve the production, transformation, and utilization of chemicals, materials, and energy. Here are some key industries where chemical engineers make significant contributions:

1. Manufacturing and Process Industries: Chemical engineers are involved in the manufacturing sector, where they develop and optimize processes for producing a wide range of products, including chemicals, pharmaceuticals, polymers, plastics, fuels, and consumer goods. They work on process design, equipment selection, and optimization to ensure efficient and cost-effective production.

2. Energy and Petrochemicals: Chemical engineers play a crucial role in the energy industry, including the production, refining, and distribution of fossil fuels. They are involved in oil and gas processing, developing technologies for clean and efficient energy production, and optimizing processes for enhanced fuel efficiency. Chemical engineers are also at the forefront of developing renewable energy sources, such as solar, wind, and biofuels.

3. Environmental Protection and Sustainability: With growing concerns about environmental impact, chemical engineers contribute to environmental

protection and sustainability efforts. They develop processes and technologies for waste management, wastewater treatment, air pollution control, and sustainable resource utilization. Chemical engineers work towards minimizing the environmental footprint of industries and developing sustainable solutions for a greener future.

4. Pharmaceutical and Biotechnology: Chemical engineers play a critical role in the pharmaceutical and biotechnology industries. They are involved in the design and optimization of processes for drug manufacturing, including formulation development, quality control, and regulatory compliance. Chemical engineers also contribute to bioprocess engineering, where they work with biological systems, such as cell cultures and enzymes, to produce therapeutic proteins and other bioproducts.

5. Food and Beverage: Chemical engineers are integral to the food and beverage industry, where they ensure the safe and efficient production of food products. They work on processes related to food processing, preservation, packaging, and quality control. Chemical engineers also contribute to developing new food products, improving nutritional value, and optimizing food safety measures.

6. Environmental and Safety Consulting: Chemical engineers provide expertise in environmental and safety consulting, helping industries comply with regulations and implement sustainable practices. They assess the environmental impact of industrial processes, develop pollution prevention strategies, and provide guidance on safety protocols and risk management.

7. Research and Development: Chemical engineers contribute to scientific research and development in academia, government laboratories, and private

industries. They work on innovative projects, such as developing new materials, exploring renewable energy sources, improving process efficiencies, and advancing technologies in various fields.

In each of these industries, chemical engineers bring their expertise in process design, optimization, safety, and sustainability to address complex challenges. Their ability to integrate scientific knowledge with engineering principles makes them invaluable contributors to the development of new technologies, the improvement of existing processes, and the advancement of industries as a whole.

Key principles and concepts in chemical engineering

Chemical engineering encompasses a wide range of principles and concepts that are fundamental to the discipline. Here are some key principles and concepts in chemical engineering:

1. Mass and Energy Balances: Mass and energy balances are fundamental concepts in chemical engineering. Mass balance involves the conservation of mass during chemical reactions and process operations, while energy balance deals with the conservation of energy. These principles are used to analyze and optimize chemical processes and ensure efficient and safe operation.

2. Thermodynamics: Thermodynamics is the study of energy and its transformations in chemical systems. Chemical engineers apply thermodynamic principles to understand and predict the behavior of chemical reactions, phase equilibria, heat transfer, and energy conversion processes. Thermodynamics is essential for designing efficient and sustainable processes.

3. Fluid Mechanics: Fluid mechanics deals with the behavior of fluids (liquids and gases) and their interaction with solid surfaces. Chemical engineers use principles of fluid mechanics to design and analyze equipment and processes involving fluid flow, such as pumps, pipelines, and reactors. Understanding fluid behavior is crucial for efficient and safe operation.

4. Heat and Mass Transfer: Heat and mass transfer are key processes in chemical engineering. Chemical engineers

study the mechanisms and rates of heat transfer (conduction, convection, and radiation) and mass transfer (diffusion, convection) to optimize processes involving heat exchange and mass transport. These principles are applied in areas such as heat exchangers, distillation columns, and drying processes.

5. Reaction Kinetics: Reaction kinetics is the study of the rates at which chemical reactions occur. Chemical engineers use kinetics principles to understand and predict reaction rates, optimize reaction conditions, and design reactors. This knowledge is crucial for developing efficient and sustainable chemical processes.

6. Process Control: Process control involves monitoring and manipulating process variables to ensure optimal operation and product quality. Chemical engineers use control systems and instrumentation to maintain desired process conditions, regulate variables, and respond to disturbances. Process control ensures safety, efficiency, and consistency in industrial processes.

7. Separation Processes: Separation processes are critical in chemical engineering, as they involve separating and purifying components of mixtures. Chemical engineers use various techniques such as distillation, absorption, extraction, and membrane processes to separate and recover desired components from complex mixtures.

8. Safety and Risk Management: Safety is a paramount concern in chemical engineering. Chemical engineers are trained to identify and mitigate risks associated with chemical processes, handle hazardous materials safely, and design systems with safety features. Risk assessment, hazard analysis, and process safety management are key concepts in ensuring safe operations.

These principles and concepts form the foundation of chemical engineering, guiding the design, optimization, and safe operation of processes and systems. Chemical engineers apply these principles to develop innovative solutions, optimize resource utilization, and ensure sustainable practices in industries ranging from manufacturing and energy to pharmaceuticals and environmental protection.

Thermodynamics

Thermodynamics is a branch of physics that deals with the study of energy and its transformations, particularly in relation to heat and work. It is a fundamental concept in chemical engineering and plays a crucial role in understanding and analyzing chemical processes and systems. Here are some key aspects of thermodynamics:

1. Laws of Thermodynamics: The laws of thermodynamics are fundamental principles that govern energy and its transformations. The four laws of thermodynamics are:

   - The Zeroth Law of Thermodynamics: If two systems are in thermal equilibrium with a third system, they are also in thermal equilibrium with each other.
   - The First Law of Thermodynamics: Energy cannot be created or destroyed, but it can be converted from one form to another. It states that the change in the internal energy of a system is equal to the heat added to the system minus the work done by the system.
   - The Second Law of Thermodynamics: The total entropy of an isolated system always increases over time. It sets the direction of natural processes and introduces the concept of irreversibility.
   - The Third Law of Thermodynamics: As the temperature approaches absolute zero, the entropy of a pure crystalline substance approaches zero.

2. Heat and Work: In thermodynamics, heat is the transfer of energy due to a temperature difference, while work is the transfer of energy due to a force acting over a distance. Both heat and work are forms of energy transfer and can cause changes in the internal energy of a system.

3. Enthalpy and Entropy: Enthalpy (H) is a thermodynamic quantity that represents the total heat content of a system at constant pressure. It is defined as the sum of the internal energy of a system and the product of pressure and volume. Entropy (S) is a measure of the disorder or randomness of a system. It quantifies the number of microscopic configurations available to a system at a given macroscopic state.

4. Thermodynamic Processes: A thermodynamic process describes the change of state of a system from one equilibrium state to another. Common types of thermodynamic processes include isothermal (constant temperature), adiabatic (no heat exchange), isobaric (constant pressure), and isochoric (constant volume) processes. Understanding these processes is essential for analyzing and designing thermodynamic systems.

5. Phase Equilibria: Phase equilibria deal with the coexistence of different phases of matter, such as solids, liquids, and gases, in a system. Thermodynamics helps in understanding phase transitions, such as melting, boiling, and condensation, and predicting the conditions at which these transitions occur.

6. Equilibrium and Spontaneity: Thermodynamics provides insights into the conditions under which a process or a reaction occurs spontaneously. It defines concepts such as equilibrium, which is a state where no further changes occur in a system, and free energy, which determines the spontaneity of a process or reaction.

7. Applications in Chemical Engineering:

Thermodynamics is applied in various areas of chemical engineering, such as process design, reactor analysis, energy systems, and separation processes. It helps in determining the energy requirements, heat transfer rates, and the efficiency of chemical processes. Thermodynamics is also crucial for the design and optimization of power plants, refrigeration systems, and environmental control systems.

In summary, thermodynamics is a fundamental concept in chemical engineering that deals with energy and its transformations. It provides a framework for understanding the behavior of chemical systems, analyzing energy transfers, and predicting the spontaneity of processes. By applying thermodynamic principles, chemical engineers can optimize processes, improve energy efficiency, and design sustainable systems.

Fluid mechanics

Fluid mechanics is a branch of physics that deals with the study of fluids, which include both liquids and gases, and their behavior under various conditions. It is a fundamental concept in chemical engineering and plays a crucial role in understanding and analyzing fluid flow, heat transfer, and other related phenomena. Here are some key aspects of fluid mechanics:

1. Fluid Properties: Fluid mechanics involves the study of fundamental properties of fluids, such as density, pressure, viscosity, and temperature. These properties affect how fluids behave and interact with their surroundings.

2. Fluid Statics: Fluid statics deals with fluids at rest and the forces acting on them. It involves studying the equilibrium of fluids and analyzing concepts such as hydrostatic pressure, buoyancy, and stability of floating bodies.

3. Fluid Dynamics: Fluid dynamics focuses on the motion of fluids and the forces that cause this motion. It includes the study of fluid flow, both in steady and unsteady states. Fluid dynamics considers concepts such as velocity, flow rate, turbulence, and boundary layers.

4. Conservation Laws: Fluid mechanics relies on the conservation laws, including the conservation of mass, momentum, and energy. These laws provide a foundation for understanding how fluids behave and how they interact with their surroundings.

5. Bernoulli's Principle: Bernoulli's principle is a

fundamental concept in fluid mechanics that relates the pressure, velocity, and elevation of a fluid in a streamline flow. It states that as the velocity of a fluid increases, its pressure decreases, and vice versa.

6. Flow Regimes: Fluid mechanics classifies flow into different regimes based on parameters such as Reynolds number, which characterizes the relative importance of inertial and viscous forces in a fluid flow. These regimes include laminar flow, turbulent flow, and transitional flow.

7. Applications in Chemical Engineering: Fluid mechanics has numerous applications in chemical engineering. It is used to analyze and design systems involving fluid flow, such as pumps, pipes, heat exchangers, and reactors. Fluid mechanics principles are crucial for optimizing process efficiency, ensuring proper mixing, and preventing fluid-related issues.

8. Computational Fluid Dynamics (CFD): With advancements in computer technology, numerical methods and simulations are widely used in fluid mechanics. CFD allows engineers to model and simulate fluid flow and heat transfer in complex systems, enabling virtual testing and optimization of designs.

Understanding fluid mechanics is essential for chemical engineers as they work with processes that involve the handling, transport, and transformation of fluids. By applying fluid mechanics principles, chemical engineers can design efficient and safe systems, optimize process performance, and overcome challenges related to fluid behavior.

Heat transfer

Heat transfer is the process of thermal energy exchange between systems or objects due to a temperature difference. It is a fundamental concept in chemical engineering and plays a vital role in various industrial processes and applications. There are three primary modes of heat transfer:

1. Conduction: Conduction is the transfer of heat through a solid material or between two solids in direct contact. It occurs due to the collision of particles within the material, which leads to the transfer of thermal energy. Heat conduction is governed by Fourier's Law, which states that the rate of heat transfer through a material is proportional to the temperature gradient across it and inversely proportional to its thermal conductivity.

2. Convection: Convection is the transfer of heat through the movement of a fluid (liquid or gas). It involves the combined effects of conduction and fluid motion. Convection can occur through natural convection, where heat transfer is driven by density differences within the fluid, or forced convection, where an external force (such as a pump or fan) is used to circulate the fluid and enhance heat transfer. Convection is important in applications such as heat exchangers, cooling systems, and fluidized bed reactors.

3. Radiation: Radiation is the transfer of heat through electromagnetic waves, without the need for a medium. It can occur in vacuum or through transparent media. All objects emit and absorb thermal radiation, with the rate of radiation dependent on their temperature

and emissivity. Radiation plays a significant role in applications such as heating, cooling, solar energy collection, and thermal insulation.

Heat transfer is a crucial consideration in chemical engineering processes, as it affects the efficiency, safety, and overall performance of systems. Chemical engineers utilize heat transfer principles in various applications, including:

- Heat Exchangers: Heat exchangers are devices used to transfer heat from one fluid to another, without them coming into direct contact. They are widely used in industries such as power generation, HVAC, and chemical processing to transfer heat between process streams and enhance energy efficiency.
- Distillation: Distillation is a separation process that involves the vaporization and condensation of components in a mixture. Heat transfer plays a critical role in providing the energy required for vaporization and condensation, allowing for the separation of components based on their boiling points.
- Reaction Engineering: Heat transfer considerations are essential in reaction engineering, where temperature control is critical to achieve desired reaction rates and product quality. Heat transfer is often required to maintain reaction temperatures, remove or supply heat to control reaction rates, and prevent undesired side reactions.
- Heat Recovery: Heat transfer techniques are employed to recover waste heat generated in various processes. This recovered heat can be reused, reducing energy consumption and improving overall process efficiency.
- Thermal Management: Heat transfer plays a crucial role in thermal management, ensuring that equipment, such as reactors, turbines, and electronics, operate within safe temperature limits. Effective heat transfer design

and strategies are essential to prevent overheating and maintain equipment integrity.

Understanding heat transfer principles is vital for chemical engineers as they design, optimize, and troubleshoot processes and systems involving temperature-sensitive operations. By applying heat transfer knowledge, engineers can improve energy efficiency, enhance process control, and ensure safe and reliable operations.

Mass transfer

Mass transfer is a fundamental concept in chemical engineering that involves the movement of components or substances from one phase to another. It plays a crucial role in various industrial processes, including separation, purification, reaction engineering, and mass transfer operations. Here are some key aspects of mass transfer:

1. Diffusion: Diffusion is the process by which molecules or particles move from an area of high concentration to an area of low concentration. It is driven by the random motion of molecules and the concentration gradient. Diffusion is essential in various processes, such as the transport of gases through porous media, the mixing of fluids, and the movement of solutes in solutions.

2. Mass Transfer Coefficients: Mass transfer coefficients characterize the rate of mass transfer in a system. They depend on various factors, including the physical properties of the substances involved, the contact area between phases, and the driving force for mass transfer (such as a concentration difference). Determining mass transfer coefficients is crucial for designing and optimizing mass transfer equipment and operations.

3. Interfacial Mass Transfer: Interfacial mass transfer occurs at the interface between two phases, such as a gas-liquid interface or a liquid-liquid interface. It involves the transfer of substances across the interface, which can happen through diffusion, convection, or both. Interfacial mass transfer is important in processes such as absorption, extraction, and distillation.

4. Mass Transfer Equipment: Mass transfer equipment is designed to facilitate efficient mass transfer between phases. Examples include packed columns, tray columns, absorption towers, and extraction equipment. These equipment designs consider factors such as surface area, contact time, and mass transfer coefficients to maximize the efficiency of mass transfer operations.

5. Separation Processes: Mass transfer plays a critical role in separation processes, where it is used to separate components or substances based on their different solubilities, volatilities, or diffusivities. Examples of separation processes that rely on mass transfer include distillation, absorption, extraction, and adsorption.

6. Mass Transfer in Reaction Engineering: Mass transfer is often a crucial factor in reaction engineering, as it affects the rate of reactant delivery to the reaction sites and the removal of products. Effective mass transfer is necessary to achieve desired reaction rates and ensure high product yields in chemical reactions.

7. Membrane Processes: Membrane processes utilize selective permeable membranes to separate components based on their size, shape, or solubility. These processes rely on mass transfer across the membrane to selectively transport specific substances while retaining others.

Understanding mass transfer principles is essential for chemical engineers, as they deal with processes that involve the movement and separation of components in various phases. By applying mass transfer knowledge, engineers can optimize separation processes, improve reaction efficiencies, and design efficient mass transfer equipment.

Reaction kinetics

Reaction kinetics is the study of the rates at which chemical reactions occur and the factors that influence these rates. It is a fundamental concept in chemical engineering and plays a crucial role in understanding and optimizing chemical reactions and reaction systems. Here are some key aspects of reaction kinetics:

1. Reaction Rate: The reaction rate represents how fast a chemical reaction proceeds with time. It is typically expressed as the change in concentration of a reactant or product per unit of time. The rate of a reaction can be determined experimentally and is influenced by factors such as reactant concentrations, temperature, pressure, catalysts, and surface area.

2. Rate Laws: Rate laws describe the mathematical relationship between the reaction rate and the concentrations of the reactants. The rate law equation is determined through experimental data and is expressed in the form of rate = $k[A]^m[B]^n$, where $[A]$ and $[B]$ are the concentrations of the reactants, k is the rate constant, and m and n are the reaction orders with respect to reactants A and B, respectively.

3. Reaction Orders: The reaction order for a reactant in the rate law equation represents the power to which the concentration of that reactant is raised. It indicates how the concentration of a reactant influences the reaction rate. Reaction orders can be zero, first, second, or even fractional values.

4. Reaction Mechanisms: Reaction mechanisms describe the step-by-step sequence of elementary reactions that

occur to convert reactants into products. Each step in the mechanism involves the collision and interaction of molecules or ions. The overall rate of the reaction is determined by the slowest step in the mechanism, known as the rate-determining step.

5. Activation Energy: Activation energy is the minimum energy required for a reaction to occur. It represents the energy barrier that must be overcome for reactant molecules to transform into products. The activation energy determines the rate of reaction and is influenced by factors such as temperature and the nature of the reactants.

6. Catalysts: Catalysts are substances that increase the rate of a chemical reaction by providing an alternative reaction pathway with lower activation energy. They are not consumed in the reaction and can enhance reaction rates and selectivity. Catalysts play a crucial role in many industrial processes, allowing reactions to proceed efficiently at lower temperatures and pressures.

7. Reaction Rate Determination: Determining the rate of a reaction experimentally involves measuring changes in concentrations or other properties of the reactants or products over time. Techniques such as spectrophotometry, chromatography, and pressure measurements can be used to determine reaction rates.

Understanding reaction kinetics is essential for chemical engineers as they design, optimize, and analyze chemical processes. By applying reaction kinetics principles, engineers can select suitable reaction conditions, design efficient reactors, control reaction rates, and improve product yields. Reaction kinetics also aids in understanding reaction mechanisms, studying the effects of variables on reaction rates, and predicting the behavior of complex reaction systems.

Process flow diagrams

Process flow diagrams (PFDs) are schematic representations of a process or system that illustrate the sequence of steps involved, the equipment and components used, and the flow of materials or information. They are commonly used in chemical engineering and other industries to visually depict the major steps and interactions in a process. Here are some key aspects of process flow diagrams:

1. Symbols and Notation: Process flow diagrams use standardized symbols and notation to represent various components, equipment, and activities. These symbols help convey information about the type of equipment, the direction of flow, and the nature of the process steps. Examples of common symbols include valves, pumps, reactors, heat exchangers, separators, and instrumentation.

2. Process Steps: A process flow diagram breaks down a complex process into individual steps or stages. Each step represents a distinct operation or unit operation that contributes to the overall process. These steps are typically represented by boxes or rectangles in the diagram, with arrows indicating the flow of materials or information between them.

3. Equipment and Components: Process flow diagrams include representations of the equipment and components used in the process. This can include vessels, reactors, columns, pumps, compressors, heaters, coolers, and other specialized equipment. Each piece of equipment is labeled and connected to the

relevant process steps through flow lines.

4. Flow of Materials or Information: Process flow diagrams illustrate the flow of materials, energy, or information through the process. This can include the movement of raw materials, intermediate products, utilities (such as water or steam), or data and instructions. The flow lines connecting the process steps and equipment indicate the direction and quantity of the flow.

5. Major Control Points: Process flow diagrams often highlight key control points or critical process parameters. These may include temperature, pressure, flow rates, composition, or other variables that need to be monitored and controlled to ensure the desired process outcome.

6. Simplified Representation: Process flow diagrams provide a simplified representation of the process, focusing on the major steps and interactions. They do not typically include detailed information about equipment sizes, specifications, or specific process conditions. Instead, they provide an overview and serve as a communication tool to convey the overall process design and operation.

Process flow diagrams are used in various stages of process design, from conceptualization and feasibility studies to detailed engineering and plant operations. They help engineers and stakeholders visualize the process, identify potential bottlenecks or inefficiencies, and communicate the process design to others involved in the project. Process flow diagrams are also helpful in safety analysis, troubleshooting, and optimization efforts.

It's important to note that process flow diagrams are just one type of diagram used in process engineering. Other diagrams, such as piping and instrumentation diagrams (P&IDs), provide more detailed information about the piping, instrumentation, and control systems associated with the process.

Equipment sizing and selection

Equipment sizing and selection is a crucial aspect of process engineering and involves determining the appropriate size and type of equipment to be used in a given process or system. It ensures that the equipment can handle the required capacity, operate efficiently, and meet the process design specifications. Here are some key considerations in equipment sizing and selection:

1. Process Requirements: The first step in equipment sizing and selection is to understand the process requirements. This includes determining the desired process parameters such as flow rate, pressure, temperature, and composition. It also involves understanding the expected variations in process conditions, as well as any specific requirements for product quality, safety, or environmental regulations.

2. Design Standards and Codes: Equipment selection should consider relevant design standards and codes, such as ASME (American Society of Mechanical Engineers) codes for pressure vessels, API (American Petroleum Institute) standards for pumps and compressors, or specific industry standards for specialized equipment. Compliance with these standards ensures that the equipment is designed, fabricated, and operated safely and meets regulatory requirements.

3. Equipment Performance: The selected equipment should have the necessary performance characteristics to meet the process requirements. This includes

considering factors such as capacity, efficiency, pressure or temperature rating, corrosion resistance, material compatibility, and reliability. Performance data from manufacturers, technical specifications, and previous experience can be used to evaluate the suitability of the equipment for the intended application.

4. Process Conditions: Equipment sizing and selection should account for the specific process conditions, including the properties of the fluids being handled, such as viscosity, density, and corrosiveness. Factors like temperature, pressure, and the presence of impurities or particulate matter may also affect equipment selection. It is important to consider potential variations and the operating range of the equipment to ensure its suitability for the full range of process conditions.

5. Maintenance and Serviceability: Consideration should be given to the maintenance and service requirements of the equipment. This includes access for maintenance activities, availability of spare parts, ease of repair, and the manufacturer's support and warranty. Choosing equipment from reputable manufacturers with a track record of reliability and good after-sales support is crucial for long-term operation and maintenance.

6. Cost Considerations: Equipment selection involves balancing technical requirements with cost considerations. The upfront cost of the equipment, installation, and associated infrastructure should be evaluated in relation to the expected performance, operational costs, and lifecycle costs. It is important to consider factors such as energy consumption, maintenance requirements, and expected equipment lifespan.

7. Integration and Compatibility: Equipment selection should also consider how the chosen equipment integrates with other process components and systems. Compatibility with existing infrastructure, availability

of utilities (such as power, water, or steam), and considerations for future expansions or modifications should be taken into account.

It is important to note that equipment sizing and selection often involve iterative processes, and it is common to consult with equipment suppliers, manufacturers, and experienced engineers to ensure optimal selection. Computer-aided design (CAD) tools, simulation software, and engineering databases can also assist in the process. Additionally, it is crucial to follow industry best practices, guidelines, and regulatory requirements while performing equipment sizing and selection to ensure safe and efficient operations.

Material and energy balances

Material and energy balances are fundamental concepts in chemical engineering and play a crucial role in process design, analysis, and optimization. They involve quantifying and tracking the flow of materials and energy in a system to ensure that the inputs and outputs are properly accounted for. Here are some key aspects of material and energy balances:

Material Balances:

1. Conservation of Mass: Material balances are based on the principle of conservation of mass, which states that the total mass of a closed system remains constant over time. This principle forms the basis for tracking the flow of materials in a process.

2. Input-Output Analysis: Material balances involve quantifying the mass of all input materials, such as feedstocks, raw materials, and utilities, as well as the mass of all output products, by-products, and wastes. The goal is to ensure that the mass of the inputs equals the mass of the outputs, accounting for any accumulation or depletion within the system.

3. Mass Flow Rates: Material balances consider the mass flow rates of each material entering and leaving the system. These flow rates can be measured or estimated based on process parameters such as flow rates, concentrations, densities, and reaction rates.

4. Stoichiometry: Stoichiometry plays a key role in material balances, especially in chemical reactions. The stoichiometric relationships between reactants and products are used to determine the mass ratios and

conversions of the involved species.

5. Assumptions and Approximations: Material balances often rely on simplifying assumptions, such as steady-state conditions (no change in system properties with time) and ideal mixing (homogeneity of materials). These assumptions help simplify the calculations but should be used with caution, considering the accuracy requirements of the specific process or system.

Energy Balances:

1. Conservation of Energy: Energy balances are based on the principle of conservation of energy, which states that energy cannot be created or destroyed, only transferred or converted from one form to another. Energy balances track the flow of energy into and out of a system.

2. Forms of Energy: Energy balances consider various forms of energy, such as heat, work, and internal energy. These forms can be transferred between the system and its surroundings through heat transfer, mechanical work, or other energy exchange mechanisms.

3. Energy Flow Rates: Similar to material balances, energy balances involve quantifying the energy flow rates into and out of the system. These flow rates can be measured or estimated based on process parameters such as temperatures, flow rates, pressure drops, and heat transfer coefficients.

4. Energy Conversion: Energy balances account for the conversion of energy from one form to another within the system. For example, in chemical reactions, energy can be released or absorbed as heat, affecting the overall energy balance.

5. Efficiency Analysis: Energy balances can be used to analyze the efficiency of energy conversion processes, such as heat exchangers, turbines, or chemical reactions.

By comparing the energy input to the energy output, efficiency metrics can be calculated to assess the effectiveness of energy utilization.

6. Heat Transfer: Heat transfer mechanisms, such as conduction, convection, and radiation, are considered in energy balances. The heat transfer rates and temperature differences between the system and its surroundings contribute to the energy balance calculations.

Material and energy balances are essential tools for chemical engineers to understand and optimize processes, evaluate system performance, troubleshoot problems, and ensure the efficient use of resources. They provide a quantitative framework for analyzing process behavior, identifying bottlenecks, and making informed decisions regarding process modifications or improvements.

Process optimization

Process optimization is a key aspect of chemical engineering that focuses on maximizing efficiency, improving performance, and achieving desired outcomes in industrial processes. It involves analyzing and modifying various parameters and variables to optimize the process conditions and achieve specific objectives, such as increasing productivity, reducing costs, improving product quality, or minimizing environmental impact. Here are some key aspects of process optimization:

1. Objectives and Constraints: Process optimization starts by defining the objectives and constraints of the system. This could include maximizing production output, minimizing energy consumption, meeting product specifications, or adhering to safety and environmental regulations. Identifying and prioritizing these objectives help guide the optimization process.

2. Data Collection and Analysis: Gathering relevant data about the process, such as process variables, operating conditions, and performance indicators, is crucial for optimization. This data can be collected through sensors, measurements, and monitoring systems. Analyzing the data helps identify areas for improvement and understand the current performance of the process.

3. Modeling and Simulation: Process modeling and simulation techniques are used to create mathematical or computational models that represent the behavior of the process. These models allow engineers to understand the process dynamics, predict the effects

of changes, and explore different scenarios without physically modifying the process. Simulation can also help identify potential bottlenecks, optimize process parameters, and test the feasibility of proposed changes.

4. Design of Experiments (DOE): DOE is a statistical method used to systematically explore the effect of different process variables on the process outcomes. It involves selecting a set of experiments with different combinations of variables and analyzing the results to determine which variables have the most significant impact on the process performance. DOE helps engineers understand the process behavior and identify optimal operating conditions.

5. Optimization Algorithms: Optimization algorithms are used to find the best set of process variables that maximize or minimize a specific objective function while satisfying the constraints. These algorithms use mathematical techniques to explore the search space and find optimal solutions. Common optimization algorithms include linear programming, nonlinear programming, genetic algorithms, and gradient-based methods.

6. Sensitivity Analysis: Sensitivity analysis involves assessing the sensitivity of the process outcomes to changes in different variables. By analyzing how variations in specific parameters affect the process performance, engineers can identify critical variables that significantly impact the outcomes. This information helps prioritize efforts for process improvement and focus on areas that offer the most significant potential gains.

7. Continuous Improvement: Process optimization is an ongoing process that requires continuous monitoring, analysis, and improvement. Once an optimal operating point is identified, it is important to establish control mechanisms and monitoring systems to maintain the

optimized conditions. Regular performance evaluations and periodic re-optimization help ensure that the process remains efficient and meets changing operational or market demands.

Process optimization is a multidisciplinary field that combines knowledge from various domains, including chemical engineering, mathematics, statistics, and computer science. It requires a systematic and data-driven approach, as well as collaboration between engineers, operators, and management to implement and sustain the optimized processes. By continuously seeking ways to enhance efficiency and performance, process optimization contributes to the competitiveness and sustainability of industrial operations.

Distillation

Distillation is a widely used separation technique in chemical engineering that involves the separation of components in a mixture based on their differences in boiling points. It is a crucial process in industries such as oil refining, petrochemicals, pharmaceuticals, and beverage production. Distillation relies on the principle that when a mixture is heated, the component with the lower boiling point vaporizes first, while the component with the higher boiling point remains in liquid form. The vapor is then condensed and collected, resulting in the separation of the components.

Here are some key aspects of distillation:

1. Distillation Columns: Distillation is typically performed in a column, which consists of a vertical vessel with trays or packing material. The column provides a large surface area for the vapor-liquid contact, facilitating the separation of the components. The design of the column depends on factors such as the composition of the mixture, desired separation efficiency, and the physical properties of the components.

2. Distillation Process: The distillation process involves two main steps: vaporization and condensation. The mixture is heated in a distillation flask or reboiler, causing the more volatile component to vaporize. The vapor rises through the column, and as it encounters cooler conditions near the top, it begins to condense. The condensed liquid is then collected as a purified product, while the remaining liquid, known as the bottoms, contains the less volatile components.

3. Tray or Packing Material: Distillation columns may have trays or packing material to enhance the vapor-liquid contact and separation efficiency. Trays are horizontal plates that allow vapor to pass through while causing liquid to accumulate and flow to the next tray. Packing materials, such as structured packing or random packing, provide a large surface area for vapor-liquid contact, promoting efficient separation.

4. Distillation Types: There are various types of distillation used in different applications. Simple distillation is suitable when the boiling points of the components are significantly different. Fractional distillation is used when the boiling points of the components are closer, and it involves multiple trays or packing sections to achieve better separation. Azeotropic distillation is employed when there is an azeotrope, which is a mixture with a constant boiling point, making separation challenging.

5. Energy Requirements: Distillation is an energy-intensive process due to the heating required to vaporize the components. Energy-efficient design and operation are important to minimize energy consumption. Techniques such as heat integration, reflux ratio optimization, and process control strategies can help improve energy efficiency in distillation processes.

6. Separation Efficiency: The separation efficiency of a distillation process is measured by the number of theoretical stages or trays required to achieve the desired separation. The higher the number of theoretical stages, the better the separation. Efficiency can be improved by optimizing operating conditions, tray or packing design, and column internals.

7. Distillation Applications: Distillation is used in a wide range of applications, including the separation of crude oil into different fractions (such as gasoline, diesel, and jet fuel) in oil refineries, the purification of solvents

and chemicals, the production of alcoholic beverages, the separation of pharmaceutical compounds, and the recovery of valuable chemicals from industrial waste streams.

Distillation is a versatile and powerful separation technique that enables the production of pure substances and valuable products from complex mixtures. The design, operation, and optimization of distillation processes are essential in achieving efficient separation, reducing energy consumption, and ensuring the quality and purity of the final products.

Extraction

Extraction is a separation process commonly used in chemical engineering to separate a desired compound or component from a mixture by selectively transferring it from one phase to another. It involves the use of a solvent or extracting agent to selectively dissolve the target compound, leaving behind the other components in the original mixture. Extraction plays a crucial role in various industries, including pharmaceuticals, food processing, environmental remediation, and petroleum refining.

Here are some key aspects of extraction:

1. Solvent Selection: The choice of solvent is crucial in extraction processes. The solvent should have a high affinity for the desired compound while being immiscible or poorly miscible with the other components of the mixture. Common solvents used in extraction include water, organic solvents like ethyl acetate or hexane, and supercritical fluids like carbon dioxide.

2. Extraction Modes: Extraction can be performed in different modes depending on the nature of the mixture and the desired outcome. Liquid-liquid extraction involves the transfer of the target compound from a liquid phase (usually an aqueous or organic phase) to another immiscible liquid phase. Solid-liquid extraction involves the transfer of the target compound from a solid matrix (such as plant material) to a liquid phase.

3. Counter-current Extraction: In counter-current extraction, multiple extraction stages are used to achieve higher efficiency. The mixture and the

extracting solvent are introduced at opposite ends of a system (such as a series of extraction columns or stages), allowing for continuous extraction and separation of the desired compound. This arrangement maximizes the concentration gradient and improves the overall extraction efficiency.

4. Partition Coefficient: The partition coefficient is a measure of the distribution of a compound between two immiscible phases. It quantifies the affinity of the compound for each phase and determines the efficiency of extraction. The partition coefficient depends on factors such as the solubility of the compound in each phase, temperature, and pH.

5. Extraction Equipment: Various equipment can be used for extraction, depending on the scale and specific requirements of the process. Common equipment includes extraction columns, continuous or batch extractors, and separators to separate the solvent and extract.

6. Solid-Liquid Extraction: Solid-liquid extraction involves the transfer of the target compound from a solid material (such as plant matter, ores, or waste) into a liquid phase. Techniques such as percolation, maceration, and Soxhlet extraction are commonly used. Solid-liquid extraction is used in applications such as herbal medicine production, extraction of metals from ores, and the extraction of flavors and fragrances.

7. Environmental Considerations: Extraction processes should consider the environmental impact and sustainability. Green extraction techniques, such as using environmentally friendly solvents, reducing solvent consumption, and optimizing energy efficiency, are being developed and implemented to minimize the environmental footprint of extraction processes.

Extraction is a versatile separation technique that allows for the isolation and purification of desired compounds from complex mixtures. Its applications range from the production of pharmaceuticals and natural products to the removal of pollutants from water and soil. The design and optimization of extraction processes involve considerations of solvent selection, process parameters, equipment design, and environmental impact, ensuring efficient and sustainable extraction operations.

Absorption

Absorption is a fundamental process in chemical engineering that involves the transfer of one or more components from a gas phase into a liquid phase. It is commonly used in various industrial applications, including gas purification, chemical separations, and environmental control. The absorption process relies on the differences in solubility or affinity of the components between the gas and liquid phases.

Here are some key aspects of absorption:

1. Gas-Liquid Contact: Absorption occurs when the gas phase and liquid phase come into intimate contact, allowing the transfer of the desired component(s) from the gas phase to the liquid phase. The efficiency of absorption depends on the surface area and contact time between the gas and liquid phases. Different contactor designs, such as packed columns or tray towers, are used to optimize the gas-liquid contact and enhance mass transfer.

2. Solvent Selection: The choice of solvent is critical in absorption processes. The solvent should have a high affinity for the component(s) to be absorbed while being immiscible or poorly miscible with the other gas components. The selection is based on factors such as solubility, chemical compatibility, stability, and cost.

3. Mass Transfer: Mass transfer is the transfer of the component(s) between the gas and liquid phases. It occurs through mechanisms such as dissolution, diffusion, and convection. The rate of mass transfer depends on factors such as the concentration gradient,

solubility, diffusivity, and interfacial area.

4. Equilibrium and Driving Force: Absorption is driven by the difference in concentration or partial pressure of the component(s) between the gas and liquid phases. At equilibrium, there is a balance between the rate of absorption and desorption. The equilibrium relationship is often described by the Henry's law or other thermodynamic models.

5. Absorption Efficiency: The efficiency of absorption is influenced by several factors, including the gas and liquid flow rates, temperature, pressure, solvent properties, and the presence of other components. Understanding these factors is important for optimizing the absorption process and achieving the desired separation or purification efficiency.

6. Regeneration: After the absorption process, the liquid containing the absorbed component(s), known as the absorbent or solvent, is often subjected to a regeneration step to recover the absorbed component(s) and reuse the solvent. Regeneration can involve processes such as heating, stripping, or distillation, depending on the nature of the absorbed component(s) and the desired application.

7. Applications: Absorption finds numerous applications in industries such as gas processing and purification, carbon capture and storage, acid gas removal, odor control, and wastewater treatment. It is also used in various chemical processes, such as the removal of impurities from process streams, separation of valuable components, and recovery of solvents or gases.

Absorption is a versatile and widely used separation process that enables the removal or recovery of specific components from gas streams. The design, optimization, and control of absorption processes require a deep understanding of mass transfer principles, thermodynamics, and fluid dynamics. By efficiently capturing desired components and controlling environmental emissions, absorption plays a critical role in meeting industry requirements, environmental regulations, and sustainable development goals.

Membrane separation

Membrane separation is a widely used separation technique in chemical engineering that involves the use of semi-permeable membranes to separate different components or species in a mixture. It is a versatile and efficient process that finds applications in various industries, including water treatment, pharmaceuticals, food and beverage production, and gas separation.

Here are some key aspects of membrane separation:

1. Membrane Types: Membranes used in membrane separation processes are typically thin films made of polymers, ceramics, or metals. The choice of membrane material depends on factors such as the desired separation performance, chemical compatibility, and operating conditions. Common membrane types include reverse osmosis (RO), nanofiltration (NF), ultrafiltration (UF), and microfiltration (MF).

2. Selectivity and Permeability: Membranes are selectively permeable, meaning they allow certain components to pass through while blocking others based on their size, shape, charge, or solubility. The selectivity and permeability of a membrane determine its separation performance and efficiency.

3. Membrane Configurations: Membrane separation processes can be performed in different configurations, including flat sheet, spiral wound, tubular, hollow fiber, and ceramic membranes. These configurations vary in terms of surface area, packing density, and flow patterns, allowing for different applications and

optimization of the separation process.

4. Membrane Separation Mechanisms: Membrane separation operates on various principles, including size exclusion, diffusion, and molecular interactions. The separation mechanism depends on the membrane type and the characteristics of the components being separated. For example, reverse osmosis relies on a pressure-driven process that separates solutes based on size and solubility, while nanofiltration uses both size exclusion and charge interactions.

5. Operating Parameters: The performance of membrane separation processes is influenced by several operating parameters, including feed pressure, temperature, flow rate, and concentration. Optimizing these parameters is crucial for achieving desired separation efficiency, productivity, and cost-effectiveness.

6. Applications: Membrane separation is used in a wide range of applications, including water desalination, wastewater treatment, removal of pollutants and contaminants, concentration and purification of food and beverage products, separation of gases in air or gas streams, and recovery of valuable components from process streams. It is also employed in biomedical applications such as blood filtration and dialysis.

7. Fouling and Cleaning: Fouling refers to the accumulation of unwanted substances or particles on the membrane surface, which can lead to reduced performance and efficiency. Cleaning methods, such as backwashing, chemical cleaning, and membrane regeneration, are employed to mitigate fouling and maintain the membrane's separation performance.

Membrane separation processes offer several advantages, including high separation efficiency, low energy consumption, scalability, and environmental sustainability. The continuous development of new membrane materials and technologies has further expanded the applications and improved the performance of membrane separation processes. With ongoing research and innovation, membrane separation continues to play a crucial role in addressing the increasing demand for efficient and sustainable separation solutions in various industries.

Types of chemical reactions

Chemical reactions can be classified into several types based on the nature of the reactants and products involved, as well as the changes that occur during the reaction. Here are some common types of chemical reactions:

1. Combination (Synthesis) Reactions: In a combination reaction, two or more reactants combine to form a single product. The general equation for a combination reaction is: $A + B \rightarrow AB$. An example is the reaction between hydrogen gas and oxygen gas to form water: $2H_2 + O_2 \rightarrow 2H_2O$.

2. Decomposition Reactions: Decomposition reactions are the opposite of combination reactions. In this type of reaction, a single compound breaks down into two or more simpler substances. The general equation for a decomposition reaction is: $AB \rightarrow A + B$. An example is the decomposition of hydrogen peroxide into water and oxygen gas: $2H_2O_2 \rightarrow 2H_2O + O_2$.

3. Displacement (Replacement) Reactions: Displacement reactions involve the exchange of atoms or groups of atoms between different reactants. There are two types of displacement reactions: single displacement and double displacement.

- Single Displacement Reactions: In a single displacement reaction, an element replaces another element in a compound. The general equation for a single displacement reaction is: $A + BC \rightarrow AC + B$. An example is the reaction between zinc and hydrochloric acid to form zinc chloride and hydrogen gas: $Zn + 2HCl \rightarrow ZnCl_2 + H_2$.

- Double Displacement Reactions: In a double displacement reaction, the positive and negative ions of two compounds switch places, resulting in the formation of two new compounds. The general equation for a double displacement reaction is: AB + CD → AD + CB. An example is the reaction between sodium chloride and silver nitrate to form sodium nitrate and silver chloride: $NaCl + AgNO_3 → NaNO_3 + AgCl$.

4. Combustion Reactions: Combustion reactions involve the rapid reaction of a substance with oxygen, usually resulting in the release of heat and light. The reactant is typically a fuel, such as a hydrocarbon, and the product is carbon dioxide and water. An example is the combustion of methane (natural gas) in the presence of oxygen: $CH_4 + 2O_2 → CO_2 + 2H_2O$.

5. Acid-Base Reactions: Acid-base reactions involve the transfer of protons (H^+) from an acid to a base. The general equation for an acid-base reaction is: Acid + Base → Salt + Water. An example is the reaction between hydrochloric acid and sodium hydroxide to form sodium chloride and water: $HCl + NaOH → NaCl + H_2O$.

These are just a few examples of the many types of chemical reactions that can occur. Chemical reactions play a fundamental role in understanding and studying the behavior of matter and are essential for various applications in chemistry, industry, and everyday life.

Reaction rate equations

Reaction rate equations describe the mathematical relationship between the rate of a chemical reaction and the concentrations of the reactants. These equations help us understand the factors that influence the speed at which a reaction occurs. Here are some common types of reaction rate equations:

1. Rate Law for Elementary Reactions: Elementary reactions are simple, single-step reactions with a defined molecularity. The rate law for an elementary reaction is determined directly from the stoichiometry of the reaction. For example, for the reaction $A + B \rightarrow C$, the rate law would be expressed as: Rate = $k[A][B]$, where $[A]$ and $[B]$ represent the concentrations of the reactants A and B, and k is the rate constant.

2. Rate Law for Complex Reactions: Complex reactions involve multiple elementary steps, and their rate laws are determined by the slowest step, also known as the rate-determining step. The rate law for a complex reaction is typically derived from experimental data. For example, if a reaction proceeds through two steps, $A + B \rightarrow X$ and $X + Y \rightarrow Z$, and the second step is the slowest, the rate law would be expressed as: Rate = $k[X][Y]$.

3. Rate Law with Fractional Order: In some cases, the rate of a reaction may not be directly proportional to the concentrations of the reactants. This can occur when the reaction follows a complex mechanism or involves intermediate species. In such cases, the rate law may have fractional orders. For example, a reaction with a rate law of Rate = $k[A]^{(1/2)}[B]^2$ has a fractional order

of 1/2 with respect to reactant A and an order of 2 with respect to reactant B.

4. Rate Law with Temperature Dependency: The rate of a chemical reaction is also influenced by temperature. The Arrhenius equation describes the temperature dependence of the rate constant (k) and is given by: $k = A * e^{(-Ea/RT)}$, where A is the pre-exponential factor, Ea is the activation energy, R is the gas constant, and T is the temperature in Kelvin. The Arrhenius equation allows us to understand how changes in temperature affect the rate of a reaction.

It's important to note that determining the rate law and rate constant for a specific reaction requires experimental data, such as measuring the change in concentration over time. Through careful analysis of this data, scientists can derive the appropriate rate equation that describes the reaction's kinetics.

Reaction rate equations provide valuable insights into the kinetics of chemical reactions and allow us to predict and control reaction rates under various conditions. They form the foundation for understanding reaction mechanisms, optimizing reaction conditions, and developing efficient industrial processes.

Catalysis and catalyst design

Catalysis is a process that involves the use of a catalyst to increase the rate of a chemical reaction without being consumed in the process. Catalysts play a crucial role in various industries by enabling more efficient and sustainable production processes. Catalyst design is the process of developing and optimizing catalysts to enhance their activity, selectivity, and stability. Here are some key aspects of catalysis and catalyst design:

1. Role of Catalysts: Catalysts provide an alternative reaction pathway with lower activation energy, allowing reactions to occur at lower temperatures and with higher reaction rates. They facilitate the breaking and formation of chemical bonds, leading to the desired products. Catalysts can increase reaction rates, improve yield, enhance selectivity, reduce energy consumption, and enable the use of more environmentally friendly reaction conditions.

2. Types of Catalysts: Catalysts can be classified into different types based on their physical and chemical properties. Some common types include heterogeneous catalysts, which exist in a different phase than the reactants (e.g., solid catalysts in a gas or liquid phase reaction), and homogeneous catalysts, which are in the same phase as the reactants (e.g., catalysts dissolved in a liquid reaction). Other types include enzyme catalysts, photo-catalysts, and biocatalysts.

3. Catalyst Design: Catalyst design involves selecting or developing materials with the desired properties to achieve specific reaction outcomes. Factors to

consider include the catalyst's surface area, pore size, composition, crystal structure, and active sites. Catalyst design may involve various approaches, such as modifying the catalyst's structure or composition, optimizing its surface properties, and incorporating additional functionalities or supports.

4. Reaction Mechanisms: Understanding the reaction mechanism is crucial for catalyst design. It involves studying the sequence of steps that occur during the reaction, including adsorption of reactants, surface reactions, desorption of products, and regeneration of the catalyst. This knowledge helps identify the key reaction intermediates, active sites on the catalyst surface, and potential limitations or bottlenecks in the reaction pathway.

5. Catalyst Characterization: Characterizing catalysts is essential for understanding their structure, composition, and activity. Techniques such as X-ray diffraction (XRD), scanning electron microscopy (SEM), transmission electron microscopy (TEM), surface area analysis, and spectroscopic methods provide valuable information about catalyst properties, including particle size, morphology, crystal structure, surface area, and chemical composition.

6. Catalyst Optimization: Catalysts can be optimized through various strategies, including doping or alloying with different elements, adjusting the catalyst support, modifying the catalyst's surface properties, and controlling the particle size or dispersion. Computational methods, such as density functional theory (DFT) calculations and molecular simulations, are also used to predict and optimize catalyst properties.

7. Applications of Catalysis: Catalysis has widespread applications in many industries, including petroleum refining, chemical production, pharmaceuticals, environmental remediation, and energy conversion.

Catalytic processes include hydrogenation, oxidation, polymerization, isomerization, and many more. Catalysis is also crucial for sustainable technologies such as renewable energy production, carbon dioxide capture, and green chemistry initiatives.

Catalysis and catalyst design play a pivotal role in advancing sustainable and efficient chemical processes. By developing innovative catalysts and understanding the underlying principles, scientists and engineers can drive advancements in various industries, reduce energy consumption and waste generation, and contribute to a more sustainable future.

Reactor design and operation

Reactor design and operation are crucial aspects of chemical engineering that involve the design, optimization, and management of chemical reactors for various industrial processes. A chemical reactor is a vessel or system where chemical reactions occur under controlled conditions. Here are some key considerations in reactor design and operation:

1. Reaction Kinetics: Understanding the kinetics of the reaction is essential in reactor design. This involves studying the rate at which the reaction occurs and determining the reaction mechanism. Reaction kinetics data, such as rate constants and reaction orders, help in selecting the appropriate reactor type and designing the optimal operating conditions.

2. Reactor Types: There are several types of reactors used in chemical processes, each with its own advantages and limitations. Common reactor types include batch reactors, continuous stirred-tank reactors (CSTR), plug-flow reactors (PFR), fixed-bed reactors, and fluidized-bed reactors. The choice of reactor type depends on factors such as reaction kinetics, reactant concentrations, heat transfer requirements, and the desired residence time.

3. Reactor Design Parameters: Reactor design involves determining key parameters such as reactor size, geometry, heat transfer requirements, catalyst loading (if applicable), and reactant feed rates. These parameters are influenced by factors such as desired conversion, reaction rate, selectivity, safety considerations, and economic considerations.

4. Heat Transfer: Heat transfer is an important aspect of reactor design, as many reactions are exothermic and require efficient cooling or heating to maintain the desired temperature. Heat transfer can be achieved through various methods, including jacketed reactors, internal heat exchangers, or external heat transfer systems.

5. Mixing and Residence Time: Achieving proper mixing is essential for homogeneous reactions and for ensuring that reactants are in contact with the catalyst (if applicable). Mixing can be achieved through mechanical agitators, baffles, or internal circulation systems. The residence time, or the time it takes for the reactants to pass through the reactor, is an important parameter in determining the reactor size and throughput.

6. Safety Considerations: Safety is of paramount importance in reactor design and operation. Designing reactors with appropriate safety features, such as pressure relief systems, temperature control, and monitoring systems, is essential to prevent accidents and ensure the well-being of personnel and the environment.

7. Reactor Modeling and Simulation: Computational modeling and simulation techniques are used to optimize reactor design and operation. Computer-aided software can simulate reaction kinetics, fluid flow, heat transfer, and other parameters to predict the behavior of the reactor under various operating conditions. This allows engineers to optimize reactor design and troubleshoot potential issues before implementation.

8. Scale-up and Commercialization: Scaling up a reactor design from the laboratory scale to the commercial scale involves considerations such as process economics, production rates, and safety regulations. Engineering principles, such as maintaining similar heat transfer rates, reaction kinetics, and mixing conditions, are

crucial in ensuring successful scale-up.

Reactor design and operation require a deep understanding of reaction kinetics, transport phenomena, thermodynamics, and safety considerations. By considering these factors and using advanced modeling and simulation tools, chemical engineers can design efficient and safe reactors for a wide range of chemical processes in industries such as petrochemicals, pharmaceuticals, polymers, and energy production.

Control system fundamentals

Control systems are an integral part of many engineering applications, enabling the automation and regulation of various processes. Control systems monitor and manipulate the behavior of dynamic systems to achieve desired objectives. Here are some key fundamentals of control systems:

1. Control Objectives: Control systems are designed to achieve specific objectives, such as maintaining a desired setpoint, tracking a reference signal, minimizing errors, optimizing performance, or ensuring system stability. These objectives vary depending on the application, whether it's controlling temperature, pressure, speed, level, or other process variables.

2. Feedback Control: Feedback control is the most common type of control system. It involves continuously measuring the system's output and comparing it to the desired reference or setpoint. Any difference between the actual output and the desired value, known as the error signal, is used to adjust the system's input or control actions to bring the output closer to the desired value. Feedback control helps maintain system stability and reduce the impact of disturbances.

3. Control Loop: A control loop is the basic structure of a feedback control system and consists of three main components: a sensor or measurement device, a controller, and an actuator or final control element. The sensor measures the system's output, the controller compares it to the setpoint, and the actuator adjusts the

system's input to bring the output closer to the desired value.

4. Proportional-Integral-Derivative (PID) Control: PID control is a widely used control technique that combines three control actions: proportional control, integral control, and derivative control. Proportional control adjusts the system's input in proportion to the error signal, integral control integrates the error over time to eliminate steady-state errors, and derivative control accounts for the rate of change of the error signal to improve system response. PID control provides a balance between stability and responsiveness.

5. Control Modes: Control systems can operate in different modes depending on the control objective and system dynamics. The most common control modes are on-off control, where the system switches between two states based on a threshold; proportional control, where the control action is proportional to the error signal; and cascade control, where multiple control loops are used in a hierarchical manner to control different aspects of the system.

6. Stability and Performance Analysis: Stability analysis is essential to ensure that a control system remains stable under various operating conditions. Stability is typically assessed by analyzing the system's transfer function or by examining the stability criteria of the control loop, such as the Nyquist or Bode stability criteria. Performance analysis involves evaluating how well the control system meets the desired control objectives, such as response time, steady-state error, overshoot, and settling time.

7. Advanced Control Techniques: Beyond PID control, various advanced control techniques are employed in more complex systems. These include model predictive control (MPC), adaptive control, fuzzy logic control, neural networks, and optimal control techniques.

These advanced techniques use mathematical models, optimization algorithms, and machine learning methods to improve control performance and address complex system dynamics.

8. Control System Design and Implementation: Designing and implementing control systems involves selecting appropriate sensors, actuators, controllers, and communication networks. It also includes tuning controller parameters, optimizing control algorithms, and integrating the control system with the overall process or system.

Control system fundamentals form the basis for understanding and implementing effective control strategies in a wide range of applications, including industrial automation, robotics, aerospace, energy systems, and many more. By applying these fundamentals, engineers can design control systems that enhance system performance, improve efficiency, and ensure reliable operation.

Feedback and feedforward control

In control systems, both feedback control and feedforward control are used to achieve desired system performance and improve the response to disturbances. Here's an explanation of feedback and feedforward control:

1. Feedback Control: Feedback control is the most common type of control system, where the system's output is continuously monitored and compared to a desired reference or setpoint. The difference between the actual output and the setpoint, known as the error signal, is fed back to adjust the system's input or control actions. The control actions are determined based on the error signal and the system's dynamics, typically using a controller such as a proportional-integral-derivative (PID) controller. By continuously monitoring the output and making adjustments based on the error, feedback control helps maintain system stability, reduce errors, and improve system performance.

2. Feedforward Control: Feedforward control, also known as anticipatory control, is a technique where the control system takes into account known disturbances or inputs that can affect the system's output. Unlike feedback control that responds to the error between the actual and desired output, feedforward control acts preemptively by adjusting the system's input based on the expected disturbance or input. This is done by measuring or estimating the disturbance or input and using a mathematical model or a predetermined relationship between the disturbance and

the system's output to determine the appropriate control action. Feedforward control helps minimize the impact of disturbances on the system's output and can significantly improve the system's response.

3. Combined Feedback and Feedforward Control: In many practical applications, both feedback and feedforward control are used together to achieve optimal system performance. Feedback control is effective in compensating for uncertainties, modeling errors, and disturbances that are not fully accounted for in the feedforward control scheme. Feedforward control, on the other hand, helps reduce the steady-state error caused by disturbances that can be measured or estimated in advance. By combining feedback and feedforward control, the system can achieve better disturbance rejection, faster response, and improved tracking of the desired setpoint.

The choice between feedback and feedforward control, or the combination of both, depends on the specific application and the characteristics of the system. In some cases, feedforward control may be sufficient to achieve the desired performance if the disturbances or inputs can be accurately measured or estimated. However, in many cases, feedback control is necessary to compensate for uncertainties, disturbances, and dynamic variations in the system. The design and implementation of feedback and feedforward control systems require a deep understanding of the system dynamics, modeling techniques, and control theory.

Overall, feedback control and feedforward control are complementary techniques that can be used in combination to achieve optimal control system performance and improve the system's response to disturbances. By effectively using both control strategies, engineers can design control systems that are robust, efficient, and capable of achieving desired system behavior.

PID controllers

PID controllers, short for Proportional-Integral-Derivative controllers, are widely used in control systems to regulate and stabilize dynamic processes. A PID controller continuously adjusts the control input based on the error signal, which is the difference between the desired setpoint and the measured process variable. The three components of a PID controller, namely proportional, integral, and derivative, work together to improve control performance and achieve desired system behavior.

1. Proportional Control: The proportional component of a PID controller responds to the present error signal and generates a control action proportional to the error. The proportional gain, represented by the parameter Kp, determines the magnitude of the control action in proportion to the error. Proportional control helps to reduce the steady-state error and improve the responsiveness of the system. However, it may lead to oscillatory behavior and steady-state errors in the presence of disturbances.

2. Integral Control: The integral component of a PID controller accumulates the past error over time and generates a control action proportional to the integral of the error. The integral gain, represented by the parameter Ki, determines the magnitude of the control action based on the integral of the error. Integral control helps to eliminate steady-state errors and enhances the system's ability to track the setpoint. However, excessive integral gain can lead to instability and overshoot in the system response.

3. Derivative Control: The derivative component of a PID controller predicts the future trend of the error based on its rate of change. It generates a control action proportional to the derivative of the error. The derivative gain, represented by the parameter Kd, determines the magnitude of the control action based on the rate of change of the error. Derivative control helps to dampen the system's response and improve stability. It is particularly effective in reducing overshoot and oscillations. However, derivative control can amplify noise in the measured signal, so proper filtering or smoothing techniques are often applied.

The PID controller combines the proportional, integral, and derivative components to generate the overall control action. The control action is calculated as the sum of the three components:

Control action = Kp * error + Ki * integral of the error + Kd * derivative of the error

The proportional component provides immediate response to the error, the integral component eliminates steady-state error, and the derivative component anticipates the system's response to changes in the error. The tuning of the PID controller involves adjusting the gains (Kp, Ki, and Kd) to achieve the desired control performance, such as stable response, fast settling time, minimal overshoot, and robust disturbance rejection.

PID controllers are widely used in various industries and applications, including process control, robotics, automotive systems, heating, ventilation, and air conditioning (HVAC), and many others. They offer a versatile and effective means of controlling dynamic processes by continuously adjusting the control action based on the feedback from the system. Proper tuning and optimization of PID controllers play a crucial role in achieving optimal control system performance.

Process optimization and advanced control strategies

Process optimization is a crucial aspect of chemical engineering and involves maximizing efficiency, improving yields, reducing costs, and ensuring safe and sustainable operation. Advanced control strategies are employed to enhance process performance and optimize various aspects of the system. Here are some key concepts related to process optimization and advanced control strategies:

1. Process Modeling: Process optimization begins with developing accurate mathematical models that describe the behavior of the system. These models can be based on first principles, empirical data, or a combination of both. The models provide insights into process dynamics, interactions, and constraints, and serve as a basis for optimization and control strategies.

2. Optimization Techniques: Optimization involves finding the best set of operating conditions or control settings that achieve the desired objectives. Various optimization techniques are employed, such as linear programming, nonlinear programming, dynamic programming, and evolutionary algorithms. These techniques consider process constraints, objectives (such as maximizing production or minimizing costs), and input/output relationships to find optimal solutions.

3. Model Predictive Control (MPC): MPC is an advanced control strategy that combines process models and

optimization algorithms. It uses the dynamic process model to predict the future behavior of the system and optimize control actions accordingly. MPC considers process constraints, objectives, and disturbances in real-time to generate optimal control strategies. MPC is particularly effective in handling multivariable systems with complex interactions.

4. Advanced Regulatory Control: In addition to PID control, advanced regulatory control strategies are employed to handle more complex control challenges. These include cascade control, adaptive control, nonlinear control, and fuzzy logic control. These strategies are designed to handle nonlinearity, uncertainties, and time-varying dynamics that may exist in the process.

5. Real-Time Optimization (RTO): RTO is a strategy that continuously adjusts operating conditions in real-time to optimize process performance. It integrates process models, optimization algorithms, and measurements to find optimal operating points based on changing conditions. RTO helps to improve energy efficiency, reduce costs, and maximize production.

6. Multivariable Control: Many industrial processes involve multiple interacting variables that need to be controlled simultaneously. Multivariable control strategies, such as decentralized control, centralized control, and coordinated control, are employed to optimize the performance of the entire system rather than individual loops. These strategies account for the interactions between different variables and ensure overall system optimization.

7. Advanced Sensing and Instrumentation: Advanced control strategies often require accurate and reliable measurements of process variables. Advanced sensing technologies, such as online analyzers, spectroscopy, and inferential sensors, are used to provide real-

time measurements of key process variables. These measurements are essential for effective control and optimization.

8. Integration of Data Analytics and Machine Learning: With the availability of large amounts of data, data analytics and machine learning techniques are increasingly used in process optimization. These techniques can analyze historical data, identify patterns, predict process behavior, and optimize control strategies. They can also help in predictive maintenance, anomaly detection, and fault diagnosis.

Process optimization and advanced control strategies play a crucial role in improving process efficiency, productivity, and safety. These strategies allow for more precise control, better resource utilization, reduced waste, and enhanced overall system performance. By employing these techniques, chemical engineers can continuously optimize processes, adapt to changing conditions, and meet the demands of modern industry.

Hazard identification and risk assessment

Hazard identification and risk assessment are important components of ensuring safety in various industries, including chemical engineering. These processes help identify potential hazards, evaluate the associated risks, and implement appropriate control measures to mitigate those risks. Here's an overview of hazard identification and risk assessment:

1. Hazard Identification: Hazard identification involves systematically identifying potential hazards in a process or workplace. This includes identifying hazardous materials, equipment, processes, and activities that have the potential to cause harm to people, property, or the environment. Hazards can include chemical hazards, physical hazards (such as fire, explosion, or mechanical hazards), biological hazards, ergonomic hazards, and psychosocial hazards. Various methods are used to identify hazards, including walkthrough inspections, review of process documentation, analysis of historical incident data, and input from subject matter experts.

2. Risk Assessment: Risk assessment involves evaluating the severity of identified hazards and the likelihood of their occurrence. The goal is to determine the level of risk associated with each hazard and prioritize them for further action. Risk assessment typically involves two main components:

a. Hazard Severity: This step involves assessing the potential consequences or harm that may arise from a hazard. It considers the potential impact on people, the environment, and assets. The severity is typically categorized as low, medium, or high based on the potential consequences.

b. Risk Likelihood: This step involves assessing the likelihood or probability of a hazard occurring. It considers factors such as the frequency of exposure, the effectiveness of existing control measures, and the potential for human error. The likelihood is typically categorized as low, medium, or high based on the probability of occurrence.

3. Risk Evaluation: Once the severity and likelihood of hazards are determined, the level of risk associated with each hazard can be evaluated. This is often done using a risk matrix or similar tool that combines the severity and likelihood assessments to assign a risk rating or level. The risk rating helps prioritize hazards based on their potential impact and guides decision-making for further risk management actions.

4. Risk Management: Risk management involves implementing control measures to mitigate or eliminate identified risks. This may include engineering controls (such as designing safer processes or installing safety systems), administrative controls (such as implementing standard operating procedures or training programs), and personal protective equipment (PPE). The control measures should be selected based on the hierarchy of controls, which prioritizes elimination or substitution of hazards before relying on other control measures. Regular monitoring, inspection, and review of control measures are also important to ensure their effectiveness and make any necessary adjustments.

5. Documentation and Communication: Hazard identification and risk assessment should be documented to maintain a record of identified hazards, risk evaluations, and control measures implemented. This information should be communicated to relevant stakeholders, including workers, management,

and regulatory authorities. Effective communication ensures that everyone is aware of the hazards, risks, and control measures in place, promoting a culture of safety and collaboration.

Hazard identification and risk assessment are ongoing processes that should be revisited regularly as new hazards are identified, processes change, or new information becomes available. By systematically identifying hazards, evaluating risks, and implementing appropriate control measures, chemical engineers can create safer work environments and reduce the likelihood of incidents and accidents.

Process safety management

Process safety management (PSM) is a comprehensive approach to identifying, understanding, and managing the risks associated with chemical processes in industries such as oil and gas, petrochemicals, and pharmaceuticals. PSM is crucial for preventing and mitigating incidents that can lead to fires, explosions, chemical releases, and other hazardous situations. Here are key aspects of process safety management:

1. Process Safety Culture: Process safety management begins with developing a strong safety culture within an organization. This involves fostering a shared commitment to safety, open communication, active participation, and a proactive approach to identifying and addressing potential hazards. A positive safety culture encourages employees at all levels to prioritize safety in their daily work.

2. Process Hazard Analysis (PHA): PHA is a systematic method for identifying and evaluating potential hazards associated with a process. It involves techniques such as Hazard and Operability Studies (HAZOP), What-If Analysis, and Failure Mode and Effects Analysis (FMEA). PHA helps identify potential deviations, failures, and vulnerabilities in the process, allowing for the implementation of appropriate safeguards and controls.

3. Operating Procedures: Well-defined and documented operating procedures are essential for ensuring safe and consistent operation of processes. These procedures outline the steps and precautions to be followed during

normal operations, startup, shutdown, and emergency situations. They provide guidance to operators and minimize the risks associated with human error.

4. Employee Training and Competency: Adequate training and competency development are crucial for ensuring that employees have the knowledge and skills to safely operate and maintain process equipment. Training programs should cover topics such as process hazards, emergency response procedures, and the proper use of safety equipment. Ongoing training and refresher courses are necessary to keep employees updated on new hazards and best practices.

5. Mechanical Integrity: Regular inspection, testing, and maintenance of equipment are vital for maintaining its integrity and preventing equipment failures that could lead to hazardous incidents. Mechanical integrity programs involve conducting inspections, testing safety systems, implementing preventive maintenance practices, and ensuring proper equipment documentation and record-keeping.

6. Management of Change (MOC): MOC processes ensure that any modifications to equipment, processes, or operating procedures are thoroughly assessed for potential safety impacts. A formal MOC process includes evaluating the risks associated with proposed changes, obtaining necessary approvals, communicating changes to affected personnel, and updating relevant documentation.

7. Emergency Planning and Response: Robust emergency planning and response systems are essential for effectively managing potential incidents. This includes developing emergency response plans, conducting drills and exercises, and providing appropriate training to employees. Emergency response plans should outline procedures for evacuations, containment of spills or releases, and communication with external emergency

response organizations.

8. Continuous Improvement and Learning: Process safety management is a continuous process of learning and improvement. Organizations should encourage a culture of continuous improvement by conducting incident investigations, analyzing near misses, sharing lessons learned, and implementing corrective actions to prevent recurrence. Regular audits and reviews of process safety management systems help identify areas for improvement and ensure compliance with applicable regulations and standards.

Process safety management is critical for protecting employees, communities, and the environment from the risks associated with chemical processes. By implementing effective PSM practices, organizations can minimize the likelihood of incidents, enhance operational reliability, and build trust with stakeholders.

Environmental regulations and sustainability

Environmental regulations and sustainability are important aspects of modern society and play a crucial role in protecting the environment, promoting sustainable development, and ensuring the well-being of future generations. Here's an overview of environmental regulations and their connection to sustainability:

1. Environmental Regulations: Environmental regulations are laws and policies implemented by governments to protect the environment, conserve natural resources, and prevent or reduce pollution. These regulations set standards and guidelines for industries, businesses, and individuals to follow in order to minimize their environmental impact. They cover a wide range of areas, including air and water quality, waste management, hazardous substances, land use, and wildlife protection.

2. Compliance and Enforcement: Environmental regulations typically require businesses and industries to obtain permits, conduct environmental impact assessments, monitor and report on their environmental performance, and implement specific pollution prevention measures. Regulatory agencies are responsible for enforcing these regulations through inspections, audits, and penalties for non-compliance. Compliance with environmental regulations is essential for preventing pollution, minimizing environmental damage, and protecting public health.

3. Sustainability: Sustainability is the concept of meeting the needs of the present generation without compromising the ability of future generations to

meet their own needs. It involves finding a balance between economic development, social well-being, and environmental protection. Sustainability encompasses various dimensions, including environmental, social, and economic considerations. It emphasizes the importance of using resources efficiently, reducing waste and emissions, promoting renewable energy, preserving biodiversity, and fostering social equity.

4. Environmental Management Systems: Many organizations adopt environmental management systems (EMS) to effectively manage their environmental impact and ensure compliance with regulations. EMS frameworks, such as ISO 14001, provide a structured approach for organizations to set environmental objectives, implement appropriate controls and procedures, monitor performance, and continually improve their environmental performance. EMS helps organizations integrate sustainability principles into their operations and foster a culture of environmental responsibility.

5. Corporate Social Responsibility: Many companies embrace the concept of corporate social responsibility (CSR), which involves voluntarily taking responsibility for the social and environmental impacts of their operations. CSR initiatives go beyond legal compliance and aim to make a positive impact on society and the environment. Companies may implement sustainable practices, support environmental conservation projects, engage in community development, promote ethical sourcing, and transparent reporting.

6. Green Technologies and Innovation: Environmental regulations and sustainability goals drive the development and adoption of green technologies. These technologies focus on reducing resource consumption, improving energy efficiency, minimizing pollution, and promoting sustainable practices.

Examples include renewable energy systems, energy-efficient technologies, waste management technologies, and green building materials. The adoption of green technologies not only helps organizations comply with regulations but also offers economic benefits, such as cost savings, improved reputation, and access to new markets.

7. International Cooperation: Environmental challenges often transcend national boundaries, and international cooperation is crucial for addressing global environmental issues. International agreements and frameworks, such as the Paris Agreement on climate change and the United Nations Sustainable Development Goals, provide a platform for countries to work together to tackle environmental challenges. These agreements promote collaboration, knowledge sharing, and the development of common strategies to achieve global sustainability objectives.

Environmental regulations and sustainability efforts are essential for creating a more sustainable and resilient future. By implementing and complying with regulations, organizations can minimize their environmental impact and contribute to the conservation of natural resources. Embracing sustainability principles goes beyond compliance and allows organizations to actively contribute to the well-being of the planet and society as a whole.

Real-world examples and case studies

Here are some real-world examples and case studies that demonstrate the connection between environmental regulations, sustainability, and their impact:

1. The Clean Air Act (United States): The Clean Air Act, implemented in the United States, has significantly improved air quality and reduced harmful emissions. This regulation sets standards for air pollutants, establishes emission limits for industries and vehicles, and promotes the use of cleaner technologies. As a result, air pollution levels have decreased, leading to improved public health and environmental conditions.

2. The European Union Emissions Trading System (EU ETS): The EU ETS is a cap-and-trade system that aims to reduce greenhouse gas emissions in the European Union. It sets a limit on the total amount of greenhouse gases that can be emitted by covered industries. Companies receive emission allowances, and if they exceed their limit, they must purchase additional allowances or invest in emission reduction projects. This system encourages companies to reduce their emissions and transition to low-carbon technologies.

3. LEED Certification: The Leadership in Energy and Environmental Design (LEED) certification is a widely recognized green building certification program. It sets criteria for energy efficiency, water conservation, indoor air quality, and sustainable site development. Buildings that meet the requirements receive LEED certification, demonstrating their commitment

to sustainable construction practices. LEED-certified buildings consume less energy, conserve water, and provide healthier indoor environments.

4. Renewable Energy Transition: Many countries and regions are implementing policies and regulations to promote the transition to renewable energy sources. For example, Germany's Renewable Energy Sources Act (EEG) provides feed-in tariffs and incentives to support the development of renewable energy projects. This has led to significant growth in renewable energy generation, reducing reliance on fossil fuels and decreasing greenhouse gas emissions.

5. Sustainable Supply Chain Practices: Companies are increasingly adopting sustainable supply chain practices to reduce their environmental impact. For example, the apparel industry has implemented initiatives to ensure responsible sourcing of materials, reduce water usage in manufacturing, and promote fair labor practices. These efforts address environmental and social issues along the entire supply chain, from raw material extraction to product disposal.

6. Waste Management and Recycling: Governments and organizations are implementing regulations and programs to promote waste management and recycling. For instance, countries like Sweden have implemented effective waste management systems that prioritize recycling and waste-to-energy conversion, resulting in minimal landfill usage. Such initiatives reduce environmental pollution, conserve resources, and promote a circular economy.

These examples highlight how environmental regulations and sustainability initiatives have led to tangible improvements in various sectors, from air and water quality to energy production and waste management. They demonstrate that by implementing and complying with regulations, businesses and communities can make a positive impact on the environment and work towards a more sustainable future.

Problem-solving exercises and simulations

Problem-solving exercises and simulations are effective tools for enhancing critical thinking, decision-making, and problem-solving skills. They provide a practical and interactive way to apply knowledge, analyze complex situations, and develop effective solutions. Here are some examples of problem-solving exercises and simulations that can be used in various fields:

1. Case Studies: Case studies present real or hypothetical scenarios that require analysis and problem-solving. Participants are presented with a situation and must identify the problem, gather relevant information, analyze the facts, and propose appropriate solutions. Case studies can be used in business, healthcare, engineering, and other fields to simulate real-world challenges and encourage critical thinking.

2. Group Discussions and Brainstorming: Group discussions and brainstorming sessions are effective problem-solving exercises that encourage collaboration and creativity. Participants discuss a problem or challenge, share ideas, and collectively generate potential solutions. This approach allows for diverse perspectives and fosters teamwork and communication skills.

3. Role-Playing Simulations: Role-playing simulations involve participants assuming different roles and acting out scenarios to solve problems. This technique is often used in management, conflict resolution, and customer service training. By stepping into different roles, participants gain insights into various perspectives and

practice problem-solving in a dynamic and realistic setting.

4. Business Simulations: Business simulations are computer-based or tabletop exercises that simulate real-world business scenarios. Participants take on the role of business managers or teams and make decisions in areas such as finance, marketing, and operations. Business simulations allow participants to experience the consequences of their decisions in a risk-free environment and develop their problem-solving and strategic thinking skills.

5. Virtual Reality (VR) Simulations: Virtual reality simulations offer immersive experiences that replicate real-world scenarios. They can be used in fields such as healthcare, aviation, and emergency response training. VR simulations allow participants to practice problem-solving in realistic and high-stakes environments, enhancing decision-making skills and preparing them for real-life situations.

6. Gaming and Gamification: Game-based learning and gamification techniques are increasingly used in problem-solving exercises. By incorporating game elements, such as points, challenges, and rewards, participants are engaged and motivated to solve problems. Gamification can be applied to various fields, including education, employee training, and organizational development.

7. Model Building and System Dynamics: Model building and system dynamics involve creating representations of complex systems to analyze and solve problems. Participants use mathematical or computer-based models to simulate the behavior of the system, identify key variables, and understand the relationships between them. This approach is commonly used in engineering, economics, and environmental sciences.

These problem-solving exercises and simulations provide a dynamic and interactive learning experience, allowing participants to apply their knowledge, develop analytical skills, and explore innovative solutions. Whether used in classrooms, training programs, or professional development workshops, these exercises offer valuable opportunities to enhance problem-solving abilities and foster critical thinking skills in a practical and engaging manner.

Application of principles to industrial processes

The application of principles to industrial processes is a fundamental aspect of chemical engineering. Chemical engineers apply their knowledge of various principles and concepts to design, optimize, and operate industrial processes in sectors such as manufacturing, pharmaceuticals, energy production, and environmental engineering. Here are some key areas where principles are applied in industrial processes:

1. Process Design: Chemical engineers apply principles such as mass and energy balances, thermodynamics, and fluid mechanics to design efficient and cost-effective industrial processes. They analyze the requirements, constraints, and objectives of the process and develop strategies to achieve desired outcomes. This includes selecting appropriate equipment, designing process flow diagrams, and ensuring safe and sustainable operation.

2. Process Optimization: Principles of process optimization, such as mathematical modeling, statistical analysis, and experimental design, are applied to improve the efficiency, productivity, and profitability of industrial processes. Chemical engineers use these principles to identify bottlenecks, optimize operating conditions, reduce waste, and enhance product quality. Optimization techniques can involve adjusting variables, implementing control strategies, and utilizing advanced technologies.

3. Reaction Engineering: The principles of reaction kinetics, catalysis, and reactor design play a vital role

in industrial processes that involve chemical reactions. Chemical engineers apply these principles to optimize reaction conditions, select suitable catalysts, design reactors with optimal performance, and maximize product yield. They also consider factors such as reaction kinetics, heat transfer, and mass transfer to ensure efficient and safe operation.

4. Separation Processes: Separation processes, such as distillation, extraction, and membrane filtration, are crucial in various industries for purifying and separating desired components from mixtures. Chemical engineers apply principles of phase equilibrium, mass transfer, and thermodynamics to design and optimize separation processes. They analyze the properties of the substances involved, determine appropriate separation techniques, and optimize operating conditions for efficient separation.

5. Process Control and Automation: Chemical engineers use principles of control systems, feedback control, and automation to maintain stable and optimal operation of industrial processes. They design and implement control strategies to monitor process variables, detect deviations, and adjust operating conditions in real-time. Process control and automation techniques enhance safety, efficiency, and productivity by minimizing variations and ensuring process stability.

6. Safety and Risk Assessment: Principles of process safety, hazard identification, and risk assessment are essential in industrial processes to ensure the safety of personnel, equipment, and the surrounding environment. Chemical engineers use these principles to identify potential hazards, assess risks, and implement safety measures. They analyze process conditions, evaluate potential consequences, and design safety systems to prevent accidents and mitigate risks.

7. Environmental Impact and Sustainability: Principles

of environmental engineering and sustainability are integrated into industrial processes to minimize environmental impact and promote sustainable practices. Chemical engineers consider factors such as waste management, energy efficiency, water conservation, and emissions control to design processes that are environmentally friendly and comply with regulations. They apply principles of environmental science, resource management, and life cycle assessment to minimize the ecological footprint of industrial operations.

By applying these principles and incorporating the latest technologies and best practices, chemical engineers contribute to the efficient, safe, and sustainable operation of industrial processes. Their expertise ensures that processes are optimized, products are produced efficiently, and environmental impact is minimized, ultimately benefiting both the industry and society as a whole.

Continuing education and professional certifications

Continuing education and professional certifications are essential for chemical engineers to stay updated with the latest advancements in the field, enhance their knowledge and skills, and demonstrate their expertise to employers and clients. Here are some key points regarding continuing education and professional certifications for chemical engineers:

1. Lifelong Learning: Chemical engineers must commit to lifelong learning to keep up with the ever-evolving nature of the field. They should actively seek opportunities for professional development, including attending conferences, workshops, seminars, and webinars. Engaging in continuous learning helps chemical engineers stay current with emerging technologies, research findings, and industry trends.

2. Advanced Degrees: Pursuing advanced degrees, such as a Master's or Ph.D. in Chemical Engineering or related fields, can provide in-depth knowledge and specialized expertise. Advanced degrees can open doors to higher-level positions, research opportunities, and academic careers. They demonstrate a commitment to professional growth and can enhance career prospects.

3. Professional Associations: Joining professional associations, such as the American Institute of Chemical Engineers (AIChE) or the Institution of Chemical Engineers (IChemE), offers access to valuable resources, networking opportunities, and

professional development programs. These associations often organize conferences, publish journals, and provide online platforms for knowledge-sharing among professionals in the field.

4. Professional Certifications: Professional certifications validate the knowledge and skills of chemical engineers and provide credibility to employers and clients. Some well-known certifications in the field of chemical engineering include the Professional Engineer (PE) license, Certified Chemical Engineer (CChE), and Certified Process Safety Professional (CCPSC). These certifications typically require a combination of education, work experience, and passing a rigorous examination.

5. Specialized Training and Workshops: Chemical engineers can benefit from specialized training and workshops focused on specific areas of interest or industry sectors. These programs provide targeted knowledge and skills development in areas such as process safety, environmental management, computational modeling, or advanced process control. Many organizations and training providers offer such programs, both in-person and online.

6. Employer-Sponsored Training: Some employers provide training programs and opportunities for their chemical engineers to enhance their skills and stay updated with industry advancements. These may include in-house training sessions, online learning platforms, or financial support for attending conferences and workshops. Chemical engineers should take advantage of such opportunities to further their professional development.

7. Continuing Education Units (CEUs): Many professional certifications require chemical engineers to earn a certain number of Continuing Education Units (CEUs) or Professional Development Hours (PDHs) to maintain

their certification status. CEUs can be obtained through attending workshops, seminars, online courses, or by participating in technical activities related to the field.

Continuing education and professional certifications demonstrate a commitment to professional growth and competence in the field of chemical engineering. They help chemical engineers stay current with advancements, expand their knowledge base, and enhance their career prospects. By investing in continuous learning and obtaining relevant certifications, chemical engineers can demonstrate their expertise and stand out in a competitive professional landscape.

Ethical considerations in chemical engineering

Ethical considerations are paramount in the field of chemical engineering, as engineers have a responsibility to protect public health, safety, and the environment. Here are some key ethical considerations in chemical engineering:

1. Health and Safety: Chemical engineers have a responsibility to prioritize the health and safety of workers, the community, and consumers. This includes designing and operating processes that minimize risks and hazards, following proper safety protocols, and ensuring compliance with regulations and industry standards.

2. Environmental Impact: Chemical engineers should consider the environmental impact of their work and strive to minimize pollution, waste generation, and resource depletion. They should explore sustainable alternatives, implement energy-efficient practices, and promote the use of green technologies and processes.

3. Public Health: Chemical engineers play a crucial role in developing products and processes that are safe for public use. They should ensure that their work adheres to strict quality control standards, complies with regulatory requirements, and promotes the well-being of consumers.

4. Ethical Use of Technology: Chemical engineers should consider the ethical implications of their work, especially when it comes to developing new technologies or products. They should assess the potential social and environmental impacts, consider

the long-term consequences, and proactively address any ethical concerns that may arise.

5. Integrity and Professionalism: Chemical engineers should uphold high ethical standards in their professional conduct. They should demonstrate honesty, integrity, and transparency in their interactions with clients, colleagues, and the public. They should also ensure that their work is free from conflicts of interest and bias.

6. Confidentiality and Intellectual Property: Chemical engineers often work with proprietary information, trade secrets, and intellectual property. They must respect confidentiality agreements, protect sensitive information, and uphold intellectual property rights.

7. Ethical Decision-Making: Chemical engineers should approach decision-making with a strong ethical framework. They should consider the potential impacts of their decisions on all stakeholders, weigh potential risks and benefits, and strive to make choices that align with ethical principles and societal values.

8. Professional Development and Lifelong Learning: Chemical engineers have an ethical obligation to stay updated with the latest developments in the field, engage in continuous learning, and adhere to professional codes of conduct. They should actively seek opportunities for professional development and stay informed about new regulations, technologies, and best practices.

By considering these ethical considerations, chemical engineers can contribute to the advancement of the field while ensuring the well-being of society and the environment. Adhering to ethical principles is essential for maintaining public trust and upholding the integrity and reputation of the profession.

Professional responsibilities and communication skills

Professional responsibilities and effective communication skills are essential for chemical engineers to succeed in their careers and contribute positively to their organizations and society. Here are some key points regarding professional responsibilities and communication skills for chemical engineers:

1. Technical Expertise: Chemical engineers have a responsibility to maintain and continually develop their technical knowledge and skills. They should stay updated with advancements in the field, seek opportunities for professional development, and strive for excellence in their work.

2. Ethical Conduct: Chemical engineers should uphold high ethical standards in their professional conduct. They should demonstrate integrity, honesty, and professionalism in all aspects of their work. This includes respecting confidentiality, avoiding conflicts of interest, and adhering to ethical guidelines and professional codes of conduct.

3. Safety and Environmental Awareness: Chemical engineers have a responsibility to prioritize safety, health, and environmental considerations in their work. They should be vigilant about potential hazards, implement proper safety protocols, and ensure that their processes and operations are environmentally sustainable.

4. Effective Communication: Communication skills are

crucial for chemical engineers to convey ideas, collaborate with colleagues, and present technical information to diverse audiences. They should be able to communicate clearly, concisely, and effectively through written reports, presentations, and interpersonal interactions. Strong communication skills enable effective teamwork, efficient problem-solving, and successful project management.

5. Teamwork and Collaboration: Chemical engineers often work in multidisciplinary teams, collaborating with professionals from different backgrounds. They should be able to work effectively in teams, respect diverse perspectives, and contribute to collective decision-making. Good teamwork skills foster innovation, creativity, and efficient project execution.

6. Adaptability and Continuous Learning: Chemical engineers operate in a dynamic field with evolving technologies and challenges. They should be adaptable and open to learning new concepts, technologies, and approaches. Embracing lifelong learning and being proactive in seeking opportunities for professional growth is essential for staying competitive and relevant in the field.

7. Leadership and Professional Growth: Chemical engineers should aspire to leadership roles and demonstrate leadership qualities. They should take initiative, show accountability, and inspire and motivate others. Additionally, they should seek opportunities for professional growth, such as pursuing advanced degrees, participating in professional organizations, and taking on challenging projects.

8. Client and Stakeholder Relationships: Chemical engineers often work with clients, stakeholders, and regulatory bodies. Building and maintaining strong relationships with these parties is crucial for successful project execution. Effective communication, active

listening, and understanding the needs and concerns of clients and stakeholders are essential skills for chemical engineers.

By embracing their professional responsibilities and honing their communication skills, chemical engineers can enhance their effectiveness, foster positive work environments, and contribute to the advancement of the field. These skills enable them to effectively communicate their technical expertise, collaborate with others, and address ethical and societal challenges in their work.

Advances in technology and materials

Advances in technology and materials have played a significant role in shaping the field of chemical engineering and have enabled new possibilities in various industries. Here are some key areas where advances in technology and materials have had a significant impact:

1. Process Automation and Control: Technological advancements, such as advanced sensors, data analytics, and automation systems, have revolutionized process control and optimization in chemical engineering. These technologies enable real-time monitoring, predictive maintenance, and improved process efficiency, leading to enhanced safety, reduced costs, and increased productivity.

2. Computational Modeling and Simulation: The availability of powerful computers and sophisticated software has transformed the way chemical engineers analyze and design processes. Computational modeling and simulation allow engineers to simulate complex systems, optimize designs, predict performance, and troubleshoot problems before implementation. This technology accelerates the development of new processes and reduces the need for costly and time-consuming experimental trials.

3. Nanotechnology and Advanced Materials: Nanotechnology has opened up new frontiers in materials science and engineering. Chemical engineers are at the forefront of developing and applying nanomaterials with unique properties for various

applications, such as energy storage, drug delivery, water treatment, and electronics. Advanced materials, including composites, polymers, and ceramics, have also been instrumental in improving product performance, durability, and sustainability.

4. Sustainable Technologies: Advances in technology have facilitated the development of sustainable processes and products in chemical engineering. From renewable energy sources and energy-efficient processes to green chemistry principles and waste minimization techniques, these technologies address environmental concerns and promote sustainable development. Chemical engineers play a crucial role in implementing and optimizing these sustainable technologies to reduce the environmental footprint of various industries.

5. Biotechnology and Bioengineering: The integration of biology and engineering has resulted in significant advancements in the field of biotechnology and bioengineering. Chemical engineers work on developing processes for the production of biofuels, pharmaceuticals, biomaterials, and bioremediation. They apply genetic engineering, metabolic engineering, and bioprocessing techniques to harness the power of microorganisms, enzymes, and biological systems for various applications.

6. Process Intensification and Miniaturization: Process intensification involves designing and developing compact, efficient, and sustainable process systems. This approach allows chemical engineers to optimize processes, reduce energy consumption, minimize waste, and enhance process safety. Additionally, miniaturization, such as microreactors and microfluidic devices, offers advantages in terms of improved control, higher selectivity, and reduced footprint.

7. Sustainable Energy Solutions: Chemical engineers have been instrumental in advancing sustainable energy

solutions. They have contributed to the development of renewable energy technologies, such as solar cells, wind turbines, and fuel cells. Additionally, they play a vital role in optimizing energy production, storage, and distribution systems to ensure efficient and sustainable energy use.

These are just a few examples of how advances in technology and materials have transformed the field of chemical engineering. By embracing these advancements, chemical engineers can develop innovative solutions, optimize processes, improve product performance, and contribute to a more sustainable and efficient future. Continuous monitoring of technological advancements and their application to real-world challenges is crucial for chemical engineers to stay at the forefront of the field.

Sustainable and green engineering practices

Sustainable and green engineering practices are vital for addressing environmental challenges, minimizing resource depletion, and promoting a more sustainable future. In the field of chemical engineering, sustainable practices aim to reduce the environmental impact of processes, products, and operations while ensuring economic viability. Here are some key aspects of sustainable and green engineering practices in chemical engineering:

1. Green Chemistry: Green chemistry focuses on the design and development of chemical processes and products that minimize the use of hazardous substances, reduce waste generation, and maximize resource efficiency. Chemical engineers play a crucial role in implementing green chemistry principles, such as using renewable feedstocks, designing safer chemicals, optimizing reaction conditions, and promoting waste prevention and recycling.

2. Energy Efficiency: Chemical engineers strive to maximize energy efficiency in process design and operation. This includes minimizing energy consumption, optimizing heat integration, and implementing energy-efficient technologies, such as heat exchangers, cogeneration, and advanced separation processes. Energy audits, process optimization, and the use of renewable energy sources are also important considerations.

3. Waste Minimization and Resource Recovery: Chemical engineers aim to minimize waste generation and

promote resource recovery through strategies like process intensification, recycling, and reuse. This involves designing processes that generate minimal waste and implementing efficient waste treatment and disposal methods. Chemical engineers also explore opportunities to recover valuable resources from waste streams, such as the extraction of chemicals or energy from by-products.

4. Life Cycle Assessment: Life cycle assessment (LCA) is a tool used by chemical engineers to evaluate the environmental impact of products and processes throughout their entire life cycle, from raw material extraction to disposal. LCA helps identify areas of improvement, assess trade-offs, and guide decision-making to minimize environmental impacts.

5. Water and Wastewater Management: Chemical engineers contribute to sustainable water management by developing processes for water treatment, desalination, and wastewater treatment. They focus on water conservation, efficient water use, and the removal of pollutants to protect water resources and ensure safe discharge into the environment.

6. Sustainable Materials: Chemical engineers work towards developing and utilizing sustainable materials, including biodegradable polymers, bio-based materials, and eco-friendly coatings. They explore alternative materials that reduce reliance on non-renewable resources, minimize environmental impact, and promote recyclability.

7. Environmental Compliance and Regulations: Chemical engineers have a responsibility to ensure compliance with environmental regulations and standards. They stay updated with evolving regulations and incorporate them into process design and operation to minimize environmental risks and maintain legal compliance.

8. Education and Outreach: Chemical engineers have

a role in educating and raising awareness about sustainable practices among colleagues, stakeholders, and the wider community. By promoting sustainable engineering principles, they can drive positive change and encourage the adoption of sustainable practices in various industries.

By integrating sustainable and green engineering practices into their work, chemical engineers can contribute to environmental stewardship, resource conservation, and the development of sustainable solutions. These practices not only benefit the environment but also promote long-term economic viability and social well-being.

Digitalization and Industry 4.0 in chemical engineering

Digitalization and Industry 4.0 have had a profound impact on the field of chemical engineering, revolutionizing processes, enhancing efficiency, and enabling new possibilities. Here are some key aspects of digitalization and Industry 4.0 in chemical engineering:

1. Data Analytics and Machine Learning: The availability of large amounts of data and advancements in data analytics and machine learning have transformed the way chemical engineers analyze and interpret data. They can now extract valuable insights from process data, optimize process parameters, predict system behavior, and detect anomalies. This enables improved decision-making, increased process efficiency, and reduced operational costs.

2. Internet of Things (IoT) and Sensors: IoT and sensor technologies have enabled the collection of real-time data from various points in chemical processes. Sensors can monitor temperature, pressure, flow rates, and other process parameters, providing valuable information for process control and optimization. IoT systems facilitate remote monitoring and control, enabling real-time adjustments and preventive maintenance.

3. Digital Twins: Digital twins are virtual replicas of physical processes or systems. Chemical engineers can create digital twins that mirror the behavior of

real-world processes, enabling them to simulate and optimize processes in a virtual environment. Digital twins allow for advanced process modeling, testing of different scenarios, and optimization of process parameters without the need for physical trials, leading to cost and time savings.

4. Process Automation and Robotics: Automation technologies have transformed process control and operation in chemical engineering. Automated systems and robotics can perform repetitive and hazardous tasks, ensuring consistent and safe operation. These technologies increase productivity, reduce human error, and enable round-the-clock operation.

5. Supply Chain Optimization: Digitalization and Industry 4.0 have enhanced supply chain management in chemical engineering. With improved connectivity and real-time data sharing, chemical engineers can optimize inventory management, track shipments, monitor quality, and streamline logistics. This leads to improved efficiency, reduced costs, and better customer satisfaction.

6. Virtual and Augmented Reality: Virtual and augmented reality technologies are increasingly being used in chemical engineering for training, process visualization, and maintenance. Virtual reality allows engineers to immerse themselves in a virtual environment and simulate different scenarios, while augmented reality overlays digital information onto the real-world environment, aiding in troubleshooting and equipment maintenance.

7. Cybersecurity: With increased digitalization comes the need for robust cybersecurity measures. Chemical engineers need to ensure the security of data, protect against cyber threats, and implement secure communication protocols. Safeguarding sensitive information, intellectual property, and operational

systems is critical to maintaining the integrity and reliability of chemical processes.

The integration of digitalization and Industry 4.0 technologies in chemical engineering has led to improved process efficiency, enhanced safety, reduced costs, and increased competitiveness. By embracing these advancements, chemical engineers can optimize processes, make data-driven decisions, and drive innovation in the field. It is important for chemical engineers to stay updated with the latest digital technologies and adapt them to their specific industry needs to fully leverage their potential benefits.

Recap of key concepts and insights

In this book, we have explored the world of chemical engineering, covering a wide range of topics and concepts. Here is a recap of the key concepts and insights covered:

1. Introduction to Chemical Engineering: We began by understanding the importance of chemical engineering in modern society and its role in various industries. We discussed how chemical engineers contribute to solving complex problems and developing sustainable solutions.

2. Definition and History of Chemical Engineering: We explored the definition and historical development of chemical engineering, tracing its roots to the Industrial Revolution and its evolution into a multidisciplinary field that combines principles of chemistry, physics, biology, and engineering.

3. Interdisciplinary Nature of the Field: Chemical engineering is an interdisciplinary field that draws upon knowledge from various disciplines to design, optimize, and operate processes that involve the transformation of raw materials into valuable products. We discussed how chemical engineers collaborate with experts in different fields to address complex challenges.

4. Role of Chemical Engineers in Various Industries: Chemical engineers play a crucial role in various industries, including petroleum and gas, pharmaceuticals, food and beverage, materials, environmental, and many more. We explored the diverse applications of chemical engineering principles

in these industries and how chemical engineers contribute to process design, optimization, and troubleshooting.

5. Key Principles and Concepts: We delved into several key principles and concepts in chemical engineering, including thermodynamics, fluid mechanics, heat transfer, mass transfer, reaction kinetics, and process optimization. These concepts form the foundation of chemical engineering and are essential for understanding and designing processes.

6. Process Equipment and Operations: We discussed different types of process equipment, such as reactors, distillation columns, and heat exchangers, and their role in chemical processes. We also explored various operations, such as mixing, separation, and purification, and their significance in process design and operation.

7. Control Systems and Safety: We examined the importance of control systems in maintaining process stability and optimizing performance. We discussed feedback and feedforward control, PID controllers, and advanced control strategies. Additionally, we covered the critical aspect of process safety, including hazard identification, risk assessment, and process safety management.

8. Sustainable and Green Engineering Practices: We emphasized the importance of sustainable and green engineering practices in minimizing environmental impact and promoting resource efficiency. We explored practices such as green chemistry, energy efficiency, waste minimization, and sustainable materials, highlighting how chemical engineers contribute to sustainability and environmental stewardship.

9. Advances in Technology and Materials: We discussed how advances in technology and materials have transformed the field of chemical engineering. Topics such as digitalization, Industry 4.0, nanotechnology,

and biotechnology were explored, showcasing how these advancements have enabled new possibilities and enhanced efficiency in chemical processes.

10. Ethical Considerations and Professional Responsibilities: We acknowledged the ethical considerations in chemical engineering, including the responsibility to prioritize safety, protect the environment, and uphold professional integrity. We also highlighted the importance of effective communication, teamwork, and continuing education in maintaining professional standards.

Throughout this book, we aimed to provide a comprehensive understanding of chemical engineering, its principles, and its applications. We hope that readers have gained valuable insights into this fascinating field and are inspired to pursue further knowledge and exploration in the world of chemical engineering.

Encouragement for further exploration and growth in the field

As you conclude this book on chemical engineering, we want to offer encouragement for further exploration and growth in the field. Chemical engineering is a dynamic and ever-evolving discipline, and there are countless opportunities for you to continue your journey of learning and discovery. Here are a few points to consider:

1. Lifelong Learning: Embrace the mindset of lifelong learning. The field of chemical engineering is constantly evolving with new technologies, methodologies, and research findings. Stay curious and open to new knowledge and developments. Attend conferences, seminars, and workshops, and engage in professional development activities to stay up-to-date with the latest advancements.

2. Specialization and Research: Consider areas of chemical engineering that pique your interest and delve deeper into those topics. You may choose to specialize in a particular subfield, such as process control, renewable energy, nanotechnology, or biotechnology. Engage in research projects to contribute to the advancement of knowledge in the field and gain practical experience.

3. Networking and Collaboration: Build a network of fellow professionals, researchers, and experts in the field. Join professional organizations, participate in industry events, and connect with colleagues to exchange ideas, share experiences, and collaborate

on projects. Networking can open doors to new opportunities, provide mentorship, and foster professional growth.

4. Industry and Practical Experience: Seek out opportunities to gain practical experience in the industry. Internships, co-op programs, and industrial placements can provide valuable insights into real-world applications of chemical engineering principles. This hands-on experience will enhance your understanding of industrial processes, operations, and challenges.

5. Embrace Sustainable Practices: With sustainability becoming increasingly important, explore ways to apply sustainable and green engineering principles in your work. Stay informed about emerging trends and practices in environmental protection, energy efficiency, and waste reduction. By incorporating sustainable practices into your work, you can contribute to a more sustainable future.

6. Teaching and Mentoring: Consider sharing your knowledge and expertise by becoming a mentor or educator. Share your experiences with aspiring chemical engineers, guide them in their learning journey, and inspire the next generation of professionals in the field. Teaching and mentoring can be fulfilling and help solidify your own understanding of the subject matter.

Remember, the field of chemical engineering offers a wide range of career opportunities in diverse industries, from energy and pharmaceuticals to food and environmental sectors. Stay passionate, adaptable, and committed to excellence. By continuously expanding your knowledge, honing your skills, and embracing new challenges, you can make meaningful contributions to the field and have a fulfilling and rewarding career in chemical engineering.

Keep exploring, stay curious, and continue to make a positive impact through your work in the world of chemical engineering. The possibilities are endless, and we wish you the very best on your continued journey of growth and success.

Reference tables and charts

Certainly! Here are a few reference tables and charts that can be useful in the field of chemical engineering:

1. Periodic Table of Elements: This table provides a comprehensive list of all known elements, organized by atomic number, symbol, and atomic weight. It is a fundamental tool for understanding the properties and behavior of elements in chemical reactions.
2. Steam Tables: Steam tables provide thermodynamic properties of water and steam at various pressures and temperatures. These tables are commonly used in the design and analysis of steam systems, such as power plants and industrial steam processes.
3. Heat Transfer Coefficient Charts: Heat transfer coefficient charts provide values of heat transfer coefficients for various types of heat exchangers, including shell-and-tube, plate, and finned-tube heat exchangers. These charts are used to estimate heat transfer rates and optimize heat exchanger design.
4. Phase Equilibrium Diagrams: Phase equilibrium diagrams, such as the pressure-temperature phase diagram for a substance, depict the regions of different phases (solid, liquid, gas) as a function of pressure and temperature. These diagrams are valuable in understanding the behavior of substances during phase transitions.
5. Conversion Factors: Conversion factor tables provide conversion factors between different units of measurement commonly used in chemical engineering,

such as mass, volume, pressure, temperature, and flow rate. These tables help in converting between different unit systems and ensuring accurate calculations.

6. Material Properties: Material property tables provide data on the physical and chemical properties of various substances, such as density, viscosity, thermal conductivity, and specific heat capacity. These tables assist in material selection, process design, and optimization.

7. Engineering Standards and Codes: Various engineering standards and codes provide guidelines and specifications for design, construction, and operation of equipment and processes in chemical engineering. These standards ensure safety, reliability, and compliance with industry best practices.

These are just a few examples of the reference tables and charts commonly used in chemical engineering. Depending on the specific area of focus, there may be additional tables and charts that are relevant. It is always important to consult reputable sources and references specific to the topic or application you are working on to ensure accuracy and reliability of the data.

Glossary of key terms

Here is a glossary of key terms commonly used in the field of chemical engineering:

1. Chemical Engineering: A branch of engineering that applies principles of chemistry, physics, biology, and mathematics to design, develop, and optimize processes for the production, transformation, and utilization of chemicals, materials, and energy.
2. Reactor: A vessel or system in which chemical reactions take place. It is designed to provide controlled conditions for reactions to occur, such as temperature, pressure, and residence time.
3. Distillation: A separation process that utilizes the differences in boiling points of components in a mixture to separate them based on their vapor-liquid equilibrium. It involves vaporizing the mixture and then condensing the vapor to obtain the desired products.
4. Mass Transfer: The movement of components from one phase to another, such as from a liquid to a gas or from a gas to a liquid. It plays a crucial role in processes like absorption, extraction, and distillation.
5. Heat Transfer: The process of transferring thermal energy from one object or substance to another through conduction, convection, or radiation. It is essential for controlling temperatures in chemical processes and ensuring efficient heat exchange.
6. Fluid Mechanics: The study of the behavior of fluids, both liquids and gases, and their interactions with forces and objects. It is crucial for understanding

fluid flow, pressure, and velocity in various process equipment and systems.

7. Thermodynamics: The study of energy and its transformation in chemical systems. It involves the analysis of properties such as temperature, pressure, and volume, and the relationships between them in determining the direction and extent of chemical reactions and processes.

8. Process Control: The application of control systems and techniques to monitor and regulate the variables in a chemical process, such as temperature, pressure, flow rate, and composition. It ensures process stability, safety, and optimal performance.

9. Reaction Kinetics: The study of the rates at which chemical reactions occur and the factors that influence them, such as temperature, pressure, concentration, and catalysts. It helps in understanding the mechanisms and behavior of chemical reactions.

10. Process Optimization: The systematic approach of improving and optimizing chemical processes to maximize efficiency, productivity, and profitability. It involves identifying and implementing changes that enhance process performance while considering economic, environmental, and safety factors.

These are just a few key terms in the field of chemical engineering. The glossary may vary depending on the specific focus and context of your work. It's always important to consult relevant resources and references for a comprehensive understanding of the terminology used in your area of interest.

Additional resources and recommended readings

Here are some additional resources and recommended readings for further exploration in the field of chemical engineering:

1. Books:
   - "Chemical Engineering: A Very Short Introduction" by David Shallcross
   - "Introduction to Chemical Engineering: Tools for Today and Tomorrow" by Kenneth A. Solen and John N. Harb
   - "Chemical Engineering Design: Principles, Practice and Economics of Plant and Process Design" by Gavin Towler and R.K. Sinnott
   - "Transport Phenomena" by R. Byron Bird, Warren E. Stewart, and Edwin N. Lightfoot
   - "Separation Process Principles" by J.D. Seader, Ernest J. Henley, and D. Keith Roper
   - "Process Dynamics and Control" by Dale E. Seborg, Thomas F. Edgar, and Duncan A. Mellichamp

2. Journals and Publications:
   - AIChE Journal: Official publication of the American Institute of Chemical Engineers, featuring research articles, reviews, and industry insights.
   - Chemical Engineering Science: A peer-reviewed journal covering a wide range of topics in chemical engineering, including

process design, modeling, and optimization.

- Industrial & Engineering Chemistry Research: Publishes cutting-edge research in areas like process control, reaction engineering, and sustainable technologies.

3. Websites and Online Resources:

- AIChE (American Institute of Chemical Engineers) website: Provides access to resources, publications, webinars, and professional development opportunities.
- Chemical Processing magazine: Offers articles, case studies, and news on the latest developments in the chemical process industries.
- Khan Academy: Provides free online courses and tutorials covering various topics in science and engineering, including chemical engineering principles.

4. Professional Organizations and Societies:

- American Institute of Chemical Engineers (AIChE): Offers networking opportunities, conferences, webinars, and access to technical resources.
- Institution of Chemical Engineers (IChemE): A global professional membership organization for chemical engineers, providing resources, publications, and training programs.
- Society of Chemical Engineers (SCE): A community of chemical engineers offering networking events, seminars, and professional development resources.

These resources can serve as valuable references and platforms for further learning, research, and professional growth in the field of chemical engineering. Remember to always consult reputable sources and stay up-to-date with the latest developments and advancements in the field.

www.ingramcontent.com/pod-product-compliance
Lightning Source LLC
Chambersburg PA
CBHW062328290526
45794CB00005B/1943